Springer Natural Hazards

More information about this series at http://www.springer.com/series/10179

Marco Scaioni
Editor

Modern Technologies for Landslide Monitoring and Prediction

Springer

Editor
Marco Scaioni
College of Surveying and Geo-Informatics
Tongji University
Shanghai
People's Republic of China

and

Department of Architecture, Built
Environment and Construction
Engineering
Politecnico di Milano
Milan
Italy

Springer Natural Hazards
ISBN 978-3-662-45930-0 ISBN 978-3-662-45931-7 (eBook)
DOI 10.1007/978-3-662-45931-7

Library of Congress Control Number: 2014957865

Springer Heidelberg New York Dordrecht London

Printed on acid-free paper

Springer-Verlag GmbH Berlin Heidelberg is part of Springer Science+Business Media
(www.springer.com)

Preface

This book entitled 'Modern Technologies for Landslide Monitoring and Prediction' was initially motivated by the need for collecting some papers presented during the 'Bilateral Workshop Italy–China on Technologies for Landslide Assessment and Prediction,' organized at Tongji University in Shanghai on 18 September 2013. In a following stage, other authors who could not attend the workshop but were carrying on relevant research in the field were invited to send their contributions. The result is a panorama on different techniques which may give support to the investigations on slope failures, focusing in particular on monitoring and prediction.

The editor wishes that this book will be useful to scientists, students, operators and officers of bodies and institutions dealing with geohazard management, politicians' as well as any other stakeholders involved, to understand the relevance of technology when coping with landslides at different levels. Without technology, such complex geo-processes cannot be thoroughly investigated and effective strategies for risk mitigation cannot be undertaken. On the other hand, technology has to be used correctly and the outputs it may provide have to be deeply analyzed and discussed. People who are experts in the use of technology and people who are experts on natural geo-processing and their modelling have to closely cooperate to make their efforts successful and more beneficial to the whole society. Consequently, this volume has also the aim of promoting cooperation between people who are dedicating their time to the Landslide Science, not only in the two countries represented here but in any regions of the world.

Shanghai, August 2014 Marco Scaioni

Acknowledgments

I would like to acknowledge first the Embassy of Italy in Beijing for supporting the organization of the 'Bilateral Workshop Italy–China on Technologies for Landslide Assessment and Prediction,' Tongji University, Shanghai (People's Republic of China), 18th September 2013. This workshop had been the occasion that moved me to work on this volume. In particular, I would like to thank the General Consul of Italy in Shanghai (Vincenzo De Luca) and the Science & Technology Councellor of the Embassy of Italy in Beijing (Prof. Plinio Innocenzi) for supporting and attending the workshop.

Acknowledgments go to the College of Surveying and Geo-Informatics of Tongji University. Also, I would like to acknowledge the Rector (Prof. Giovanni Azzone) and vice-Rector (Prof. Giuliano Noci) of Politecnico di Milano (Italy) for their kind support.

Many thanks go to the invited speakers from China and Italy, and to colleagues and students in the College of Surveying and Geo-Informatics (Tongji University) who helped prepare and run the workshop. Great thanks go to Lian Cao who coordinated the operative organization of the workshop.

Acknowledgments also go to 863 National High-tech R&D Program of China (No. 2012AA121302), to 973 National Basic Research Program of China (No. 2013CB733204), and to the International Cooperation Office of Tongji University.

Scientific Committee

Diego Arosio (Politecnico di Milano, Italy)
Timo Balz (Wuhan University, People's Republic of China)
Luigi Barazzetti (Politecnico di Milano, Italy)
Maurizio Barbarella (University of Bologna, Italy)
Alessandro Capra (University of Modena and Reggio Emilia, Italy)
Nicola Casagli (University of Florence, Italy)
Alessandro Ferretti (Telerilevamento Europa, Italy)
Yang Hong (Oklahoma University, US)
Runqiu Huang (Chengdu Technical University, People's Republic of China)
Yu Huang (Tongji University, People's Republic of China)
Gregorio Mannucci (Regione Lombardia, Italy)
Mingsheng Liao (Wuhan University, People's Republic of China)
Chun Liu (Tongji University, People's Republic of China)
Jianguo Liu (Imperial College London, UK)
Laura Longoni (Politecnico di Milano, Italy)
Ping Lu (Tongji University, People's Republic of China)
Zhang Lu (Wuhan University, People's Republic of China)
Monica Papini (Politecnico di Milano, Italy)
Daniele Perissin (Purdue University, US)
Francesco Pirotti (University of Padua, Italy)
Mattia Previtali (Politecnico di Milano, Italy)
Gang Qiao (Tongji University, People's Republic of China)
Riccardo Roncella (University of Parma, Italy)
Marco Scaioni (Tongji University, People's Republic of China)
Filiz Sunar (Istanbul Technical University, Turkey)
Benni Thiebes (University of Vienna, Austria)
Luigi Zanzi (Politecnico di Milano, Italy)
Antonio Vettore (University of Padua, Italy)
Xiaohua Tong (Tongji University, People's Republic of China)

Contents

Part II Geophysical and Geotechnical Methods

Part III Satellite Remote-Sensing Techniques

Introduction

Marco Scaioni

1 Landslide Technology in China and Italy

Different technologies have been largely supporting the landslide science in recent years, with the aim of achieving a better knowledge of Earth phenomena, of reducing the landslide risk, and of improving disaster mitigation and preparedness capability.

Technology, including both hardware and software development, gives a core contribution to almost all the sub-domains of landslide science. According to the classification adopted during the last World Landslide Forum (see Sassa et al. 2014), in *physical modelling* and *material testing,* it provide adequate infrastructures for the characterization of geotechnical properties of different types of soil and rock under solicitation of main triggering factors such as rainfalls and earthquakes. *Numerical modelling* helps understand the processes leading to slope failure and their magnitude (e.g., landslide mass and runoff distance), often involving severe computational aspects besides the merely scientific issues. Moreover, integration and interaction between data observation and modelling are becoming more and more important. *Remote sensing* gives today an unprecedented knowledge of Earth processes thanks to a continuously growing constellation of multi-sensors satellites, complemented by airborne and ground-based techniques (see a recent review in Scaioni et al. 2014b). Almost any spaceborne sensors available offer its contribution to the landslide science: optical and microwave imaging systems, laser scanning, and rainfall measurement missions (Tofani et al. 2014). Moreover, remote sensing can be considered as a transversal domain in the landslide science, covering almost

M. Scaioni (✉)
College of Surveying and Geo-Informatics, Tongji University, 1239 Siping Road, Shanghai 200092, People's Republic of China
e-mail: marco.scaioni@polimi.it

M. Scaioni
Department of Architecture, Built Environment and Construction Engineering, Politecnico di Milano, via Ponzio 31, 20133 Milan, Italy

M. Scaioni (ed.), *Modern Technologies for Landslide Monitoring and Prediction*, Springer Natural Hazards, DOI 10.1007/978-3-662-45931-7_1

all tasks of investigations, as they have been traditionally classified after Mantovani et al. (1996): landslide inventory and database mapping, landslide monitoring, and landslide hazard assessment and prediction. *Monitoring* the evolution of an unstable slope generally may provide important information to achieve a better knowledge of the active processes for *prediction* and *early warning* of potential geo-disasters. The integration of the observed landslide displacements, soil properties, and rainfall measurement into predictive models is the common approach towards prediction of landslides (see Arbanas et al. 2014a). Here, the technological development offered new chances to record more information from the field. While on one side remote sensing plays a growing role in landslide monitoring, ground-based methods (geodetic, extensometer, global navigation satellite system techniques) and subsurface (geotechnical and hydrogeological sensors in boreholes, ground-penetrating radar, microseismic sensors) are still fundamental to give locally precise in situ observations. An important trend can be noticed on the development of *smart sensor networks*, *distributed sensors* (such as MEMS—micro-electro-mechanical systems), and in general *low-cost* solutions (see, e.g., Eyo et al. 2014). These are of paramount relevance for the implementation in the developing countries and to face the budgetary limitations which more and more often are faced in the developed countries as well. In *landslide risk assessment*, the technology contributes in a kind of indirect way, i.e., by providing information about predisposing factors (geomorphology, topography, geology, land cover, and hydrogeology), triggering factors (mainly rainfall and earthquakes), and for model validation. Here, remote sensing techniques have become more and more important in the general context of spatial geo-information, especially for the assessment at regional level (Van Westen et al. 2008). According to Nadim (2014), the support of technology in the field of *strategical risk reduction* is at both non-structural and structural measures. In the former, it contributes to the preparation of land use plans and to early-warning systems, which can be considered as a tool for risk mitigation. Also, the development of *decision support systems* is an important contribution to this field (see Grande et al. 2008; Nadim et al. 2014). In the latter, the technology is strictly related to the design and construction of physical protection barriers. Due to their peculiarity, the structural measures are out of the scope of this book. The last subdomain where up-to-date technology has allowed an important step forward, especially with satellite remotely sensed data, is landslide mapping for inventory and database creation and update (see Arbanas et al 2014b; Guzzetti et al. 2012).

This book collects 12 chapters describing the application of innovative techniques for different stages of landslide investigation (see Sect. 2.1). The main attention is pinpointed to monitoring and prediction. About a half the papers have been presented during a bilateral workshop between Chinese and Italian researchers held in Shanghai on September 2013, as highlighted in the preface. This is the motivation of the nationality of the chapters' authors in the book.

Both countries involved, People Republic of China and Italy, have been largely affected by problems related to geo-hazards and especially by slope failures. In the major Italian poem written by Dante Alighieri (1265–1321 AD), the *Comedy*, he mentions a great rockslide (the '*great ruina*') you could come across when travelling from the cities of Trento and Verona, which probably should have been quite well known in the XIV Century.[1] It is interesting to notice that he also describes the possible causative reasons of that landslide, i.e., an earthquake or sinking of the ground, proving that some basic concepts of landslide science were already known. But several large and disruptive landslides have happened in the last decades as well, like the Vajont landslide (Southern Alps, 1963) killing more than 1,900 people as a consequence of the tsunami produced by the sudden fall of the rock mass in the water of an artificial basin (Del Ventisette et al. 2015), or the Val Pola landslides (Azzoni et al. 1992) which is also the subject of chapter A New Approach to Satellite Time Series Co-Registration for Landslide Monitoring in this book. On the Chinese side, we limit to mention to the unprecedented number of geo-disasters that followed the Wenchuan earthquake in Sichuan province on 12th May 2008: Several types of landslides were triggered due to severe seismic solicitations during a period of intense precipitations, resulting in catastrophic secondary effects such as river damming and subsequent floods (Yin et al 2009; Huang et al. 2012; Huang and Fan 2013). The response of the scientific communities in both countries resulted in an intense activity by individuals, teams, and institutions. By looking at a statistics published in Gokceoglu and Sezer (2009) on the international literature on landslides, the author who had published more papers in recognized scientific journals is a Chinese, while Italy scored 11 scholars among the best 44 in the same ranking list. Although I would not like to classify the scientific value on the mere basis of rankings, these results tell quite clearly how much is the effort put by Chinese and Italian people to develop the landslide science. As further confirmation, locations of the latest World Landslide Forums were just in Italy (Rome, 2011) and China (Beijing, 2014).

2 Content of this Volume

The book is organized in three Parts. In the first Part, five chapters deal with the topic of 'ground-based monitoring techniques'. A large relevance is given to so-called area-based deformation measurement (ABDM) methods implemented with *ground-based* (or *terrestrial*) *remote-sensing techniques*: *digital photogrammetry* (chapters Close-Range Photogrammetric Techniques for Deformation Measurement: Applications to Landslides and A Fixed Terrestrial Photogrammetric System for Landslide Monitoring), terrestrial laser scanning (TLS) (chapters A New Approach Based on Terrestrial Remote Sensing Techniques for Rock Fall Hazard Assessment and Multi-temporal Terrestrial Laser Scanning Survey of a Landslide), *ground-based interferometric synthetic aperture radar* (chapter A New Approach Based on Terrestrial Remote Sensing Techniques for Rock Fall Hazard Assessment), *infrared*

[1] Comedy, Inferno XII, 4–9: *Qual è quella ruina che nel fianco/di qua da Trento l'Adice percosse,/o per tremoto o per sostegno manco,/* … (As that landslide that on the southern side of Trento run-out on the Adige river, owing to earthquake or lack of support at the toe, …).

thermography (chapter A New Approach Based on Terrestrial Remote Sensing Techniques for Rock Fall Hazard Assessment), and their integration. Global navigation satellite systems (GNSS) networks are discussed in chapters Multi-temporal Terrestrial Laser Scanning Survey of a Landslide and Micro-scale Landslide Displacements Detection Using Bayesian Methods Applied to GNSS Data.

Part 2 contains three chapters (Analysis of Microseismic Activity Within Unstable Rock Slopes, The State of the Art of SPH Modeling for Flow-slide Propagation and Predictability of a Physically Based Model for Rainfall-induced Shallow Landslides: Model Development and Case Studies) on 'geophysical and geotechnical methods'. Here, different technologies and numerical methods are applied to predict slope failures on the basis of input observations.

Satellite remote-sensing techniques are focused in Part 3, which is composed of three chapters. Interferometric synthetic aperture radar (InSAR) techniques are discussed in chapters Monitoring Landslide Activities in the Three Gorges Area with Multifrequency Satellite SAR Datasets and Radar Technologies for Landslide Detection, Monitoring, Early Warning and Emergency Management. Chapter A New Approach to Satellite Time Series Co-Registration for Landslide Monitoring deals with the problem of geometric co-registration of satellite time series of images.

2.1 Ground-Based Monitoring Techniques

The chapter Close-Range Photogrammetric Techniques for Deformation Measurement: Applications to Landslides (Scaioni et al. 2014a) gives an introduction to *close-range photogrammetry* for the measurement of deformations on landslides. This approach presents several important advantages, such as the non-contact operational capability, the large slope coverage, the high degree of automation, the high acquisition rate, the chance to derive information on the whole surface thanks to ABDM methods, not limited to a few control points, and, generally, a lower cost with respect to 3D scanning technology. Applications are organized into two categories: (1) surface point tracking, and (2) comparison of surfaces obtained from dense image matching. Different camera configurations and geometric models to transform points from the image space to the object space are also discussed. In the last part of the Chapter, a review of the literature and two case studies are reported.

The design and application of a *photogrammetric system* made up of two fixed single-lens reflex cameras for long-term monitoring of landslides is presented in chapter A Fixed Terrestrial Photogrammetric System for Landslide Monitoring (Roncella and Forlani 2014). This chapter shows the first results of such a stereo-camera system developed to monitor shape changes of the Mont de la Saxe landslide, Courmayeur (Italy). Images are periodically acquired and sent to a remote computer station. Once an image pair is received from both cameras, a digital surface model (DSM) of the slope is generated by digital image correlation (see, e.g., Previtali et al. 2014) and made available for archiving or successive analysis.

First comparisons with results of a ground-based InSAR (GBSAR, see Monserrat et al. 2014) system have shown a good agreement.

In chapter A New Approach Based on Terrestrial Remote Sensing Techniques for Rock Fall Hazard Assessment (Mazzanti et al. 2014), the *combination* of TLS, *GBSAR*, and *infrared thermography* is used to identify both predisposing factors (mapping of discontinuities) and state of activity indicators of rock instabilities in coastal cliffs. These outputs are used for the creation of a *prioritization map* of the cliff in terms of necessary stability interventions. Owing to the fact that vertical rock faces are generally difficult to be investigated because of the limited accessibility, terrestrial remote-sensing techniques represent a great opportunity to cope with them. The application to a real coastal cliff located in southern Italy is reported in the chapter.

Chapter Multi-temporal Terrestrial Laser Scanning Survey of a Landslide (Barbarella et al. 2014) presents the methodology adopted for TLS deformation measurement of a large slump landslide in Campania region (southern Italy). In particular, two aspects of the chapter deserve particularly attention. The first one concerns the reference datum setup based on a GNSS network. The second one is the evaluation of the quality of digital terrain models (DTM's) obtained from TLS surveying over the entire region of interest. To detect the kinematics of the landslide in its several parts, both the DTM and profiles have been used, solution that have proven to be particularly useful for the interpretation of details. After the localization of various landslide bodies, the evaluation of the volumes mobilized over time has been carried out by differencing the DTM's obtained at different observation epochs.

In chapter Micro-scale Landslide Displacements Detection Using Bayesian Methods Applied to GNSS Data (Pirotti et al. 2014), a GNSS network was installed to measure the displacements of six control points near a landslide body. In particular, the chapter focuses on the methodology for the statistical analysis of point displacements, which is necessary to scrutinize between significant changes and measurement noise. Due to the small size of expected displacements, very close to their uncertainty, standard statistical testing (Teunissen 2000) may not be able to detect them. An approach based on *Bayesian inference* was applied to identify the areas on the ground with statistically significant vertical (downwards) shifts. This application shows that the Bayesian approach can be considered for the integration into statistical significance testing procedures of geodetic networks for deformation measurement (Betti et al. 2001).

2.2 Geophysical and Geotechnical Methods

Chapter Analysis of Microseismic Activity Within Unstable Rock Slopes (Arosio et al. 2014) presents the concept of *passive seismics* as a method for monitoring the propagation of cracks within a rock mass as a result of load stress or water freezing. The aim is to assess the possibility of recording precursory signals of rock mass failures on the basis of a microseismic sensor network to be used as early-warning alert. A test field where cracks were artificially induced in a rock mass was used in a first stage to demonstrate the existence of precursory patterns in the recorded observations and to characterize their properties. Then, an application to a real rock face suffering from frequent rock fall events (Mt. San Martino, Lecco, Italy) is reported. Some results related to the first months of serviceability are shown and analysed in the chapter.

A review of the state of the art of modelling flow slides on the basis of smoothed particle hydrodynamics (SPH) method is addressed in chapter The State of the Art of SPH Modeling for Flow-slide Propagation (Dai and Huang 2014). Flow slides, landslide-generated waves, and debris flows are recurring problems in mountainous areas, with high disruptive potential due to the long runoff. Predicting their propagation stage is consequently of great importance for disaster mitigation. The SPH method is a mesh-free particle technique which has been widely applied for modelling flow-slide evolution with some success.

In chapter Predictability of a Physically Based Model for Rainfall-induced Shallow Landslides: Model Development and Case Studies (Hong et al. 2014), a cost-effective physical model (SLope-Infiltration-Distributed Equilibrium—SLIDE) to identify the spatial and temporal occurrences of rainfall-induced landslides is presented. Remote-sensing observations on the rainfall rate, a DTM, a land cover map, and in situ data were adopted. The main feature of SLIDE is that it takes into account of some simplified hypotheses on water infiltration and defines a direct relationship between the factor of safety and the rainfall depth on an infinite slope model. This prototype has been applied to two case studies in Indonesia and Honduras during heavy rainfall events brought by typhoon and hurricane, respectively. Simulation results from SLIDE demonstrated good skills in predicting rainfall-induced shallow landslides by assimilating the most important dynamic triggering factors (i.e., rainfall) quantitatively. The model's prediction performance also suggested that SLIDE could serve as a potential tool for the future landslide early-warning system.

2.3 Satellite Remote-Sensing Techniques

In chapter Monitoring Landslide Activities in the Three Gorges Area with Multifrequency Satellite SAR Datasets (Zhang et al. 2014), the applications of multiple methods based on InSAR data processing for landslide monitoring in the Three Gorges area (Chongqing Municipality/Hubei Province, P.R. China) is presented. Multifrequency satellite SAR data sets acquired by ENVISAT ASAR, ALOS PALSAR, and TerraSAR-X from different orbits were analysed to retrieve historic deformations of a few typical landslides. The experimental results suggested that *advanced differential InSAR* methods (Crosetto et al. 2005) can be effectively used to monitor slow-moving landslides, while *pixel-offset tracking* method is more suitable for detecting deformation of fast-moving landslides.

Furthermore, qualitative correlation analyses indicated that variation of reservoir water level, particularly the rapid water level decrease due to discharge should be identified as a key driving factor for landslide deformation in the Three Gorges area.

In chapter Radar Technologies for Landslide Detection, Monitoring, Early Warning and Emergency Management (Del Ventisette et al. 2014), different applications of InSAR techniques are presented to prove their applicability for landslide investigation. Both spaceborne (operating in C-band) and ground-based (operating in Ku-band) sensors have been adopted. In particular, four applications in different Italian regions are shown and discussed: *landslide detection* (Arno river basin), *landslide monitoring* (Santo Stefano d'Aveto), *landslide hazard assessment* (Montebeni landslide), and *landslide emergency management* (Santa Trada landslide). Such real cases highlighted the capability and suitability of these techniques to work in different operative settings (i.e., different phenomena and geological framework) and for different aims (survey, early warning, and emergency assessment).

Image-to-image *co-registration* is one of the preprocessing steps needed for the analysis of satellite time series in both optical and microwave domains. Chapter A New Approach to Satellite Time Series Co-Registration for Landslide Monitoring (Barazzetti et al. 2014) presents a new approach where all the available images are simultaneously co-registered, overcoming the limits of traditional techniques which work on a pairwise basis. This method was tested on the flood and landslide that occurred in Valtellina (northern Italy) during summer of 1987, resulting in the large rockslide of Val Pola (Azzoni et al. 1992). A data set made up of medium-resolution satellite images collected with Landsat-4 and Landsat-5 Thematic Mapper over a period of 30 years was automatically processed. Results showed that the new approach can provide sub-pixel accuracy close to manual measurements, which today are considered the most accurate method for image registration. The multi-image co-registration method also demonstrated to be atmospheric resistant and robust against land cover changes, snow, and cloud cover.

References

Arbanas, Ž., Fathani, T. F., Shoaei, Z., Chae, B. G., & Tommasi, P. (2014a). Introduction: Monitoring, prediction and warning of landslides. In K. Sassa et al. (Eds.), *Landslide science for a safer geoenvironment* (Vol. 2, pp. 539–544). Berlin, Heidelberg: Springer.

Arbanas, S. M., Komac, M., Gokceoglu, C., & Paulin G. L. (2014b). Introduction: Landslide inventories and databases. In K. Sassa et al. (Eds.), *Landslide science for a safer geoenvironment* (Vol. 2, pp. 783–785). Berlin, Heidelberg: Springer.

Arosio, D., Longoni, L., Papini, M., & Zanzi, L. (2014). Analysis of microseismic activity within unstable rock slopes. In M. Scaioni (Ed.), *Modern technologies for landslide investigation and prediction* (pp. 141–154). Berlin, Heidelberg: Springer.

Azzoni, A., Chiesa, S., Frassoni, A., & Govi, M. (1992). The Valpola landslide. *Engineering Geology, 33*(1), 59–70.

Barazzetti, L., Gianinetto, M., & Scaioni, M. (2014). A new approach to satellite time series co-registration for landslide monitoring. In M. Scaioni (Ed.), *Modern technologies for landslide investigation and prediction* (pp. 233–249). Berlin, Heidelberg: Springer.

Barbarella, M., Fiani, M., & Lugli, A. (2014). Multi-temporal terrestrial laser scanning survey of a landslide. In M. Scaioni (Ed.), *Modern technologies for landslide investigation and prediction* (pp. 89–121). Berlin, Heidelberg: Springer.

Betti, B., Sansò, F., & Crespi, M. (2001). Deformation detection according to a Bayesian approach. In B. Benciolini (Ed.), *IV Hotine-Marussi Symposium on Mathematical Geodesy, IAG Symposia* (Vol. 122, pp. 83–88). Berlin, Heidelberg: Springer. doi:10.1007/978-3-642-56677-6_12.

Crosetto, M., Crippa, B., Biescas, E., Monserrat, O., Agudo, M., & Fernández, P. (2005). State-of-the-art of land deformation monitoring using SAR interferometry. *Photogrammetrie Fernerkundung Geoinformation, 6*, 497–510.

Dai, Z., & Huang, Y. (2014). The state of the art of SPH modelling for flow-slide propagation. In M. Scaioni (Ed.), *Modern technologies for landslide investigation and prediction* (pp. 155–164). Berlin, Heidelberg: Springer.

Del Ventisette, C., Gigli, G., Bonini, M., Corti, G., Montanari, D., Santoro, S., et al. (2015). Insights from analogue modelling into the deformation mechanism of the Vaiont landslide. *Geomorphology, 228*, 52–59. doi:10.1016/j.geomorph.2014.08.024.

Del Ventisette, C., Gigli, G., Tofani, V., Lu, P., & Casagli, N. (2014). Radar technologies for landslide detection, monitoring, early warning and emergency management. In M. Scaioni (Ed.), *Modern technologies for landslide investigation and prediction* (pp. 209–232). Berlin, Heidelberg: Springer.

Eyo, E., Musa, T., Omar, K., Idris, KM., Bayrak, T., & Onuigbo, I. (2014). Application of low-cost tools and techniques for landslide monitoring. In K. Sassa et al. (Eds.), *Landslide science for a safer geoenvironment* (Vol. 2, pp. 557–1562). Berlin, Heidelberg: Springer.

Gokceoglu, C., & Sezer, E. (2009). A statistical assessment on international landslide literature (1945–2008). *Landslides, 6*, 345–352.

Grande, O., Trucco, P., Longoni, L., & Papini, M. (2008). A Bayesian-based decision support tool for assessing and managing rock fall disasters. In *Proceedings of the International Conference on Building Education and Research: Building Resilience (BEAR 2008)*'. Heritance Kandalama, Sri Lanka, February 11–15, 2008.

Guzzetti, F., Mondini, A. C., Cardinali, M., Fiorucci, F., Santangelo, M., & Chang, K. T. (2012). Landslide inventory maps: New tools for an old problem. *Earth Science Reviews, 112*, 42–66.

Hong, Y., He, X., Cerato, A., Zhang, K., Hong, Z., & Liao, Z. (2014). Predictability of a physically-based model for rainfall-induced shallow landslides: Model development and case studies. In M. Scaioni (Ed.), *Modern technologies for landslide investigation and prediction* (pp. 165–178). Berlin, Heidelberg: Springer.

Huang, R. Q., & Fan, X. M. (2013). The landslide story. *Nature Geoscience, 6*, 325–326.

Huang, R. Q., Pei, X. J., Fan, X. M., Zhang, W. F., Li, S. G., & Li, B. L. (2012). The characteristics and failure mechanism of the largest landslide triggered by the Wenchuan earthquake, May 12, 2008, China. *Landslides, 9*, 131–142.

Mantovani, F., Soeters, R., & van Westen, C. J. (1996). Remote sensing techniques for landslide studies and hazard zonation in Europe. *Geomorphology, 15*(2), 213–225.

Mazzanti, P., Brunetti, A., & Bretschneider, A. (2014). A new approach based on terrestrial remote-sensing techniques for rock fall hazard assessment. In M. Scaioni (Ed.), *Modern technologies for landslide investigation and prediction* (pp. 69–87). Berlin, Heidelberg: Springer.

Monserrat, O., Crosetto, M., & Luzi, G. (2014). A review of ground-based SAR interferometry for deformation measurement. *ISPRS Journal of Photogrammetry and Remote Sensing, 93*, 40–48.

Nadim, F., Kalsnes, B., Solheim, A. (2014). Plenary: Progress of living with landslide risk in Europe. In K. Sassa et al. (Eds.), *Landslide science for a safer geoenvironment* (Vol. 1, pp. 3–20). Berlin, Heidelberg: Springer.

Pirotti, F., Guarnieri, A., Masiero, A., Gregoretti, C., Degetto, M., & Vettore, A. (2014). Micro-scale landslide displacements detection using Bayesian methods applied to GNSS data. In M. Scaioni (Ed.), *Modern technologies for landslide investigation and prediction* (pp. 123–138). Berlin, Heidelberg: Springer.

Previtali, M., Barazzetti, L., & Scaioni, M. (2014). Accurate 3D surface measurement of mountain slopes through a fully automated imaged-based technique. *Earth Science Informatics, 7*(2), 109–122.

Roncella, R., & Forlani, G. (2014). A fixed terrestrial photogrammetric system for landslide monitoring. In M. Scaioni (Ed.), *Modern technologies for landslide investigation and prediction* (pp. 43–67). Berlin, Heidelberg: Springer.

Sassa, K., Canuti, P., & Yin, Y. (2014). *Landslide science for a safer geoenvironment* (Vol. 2). Berlin, Heidelberg: Springer.

Scaioni, M., Feng, Y., Lu, P., Qiao, G., Tong, X., Li, R., et al. (2014a). Close-range photogrammetric techniques for deformation measurement: Applications to landslides. In M. Scaioni (Ed.), *Modern technologies for landslide investigation and prediction* (pp. 13–41). Berlin, Heidelberg: Springer.

Scaioni, M., Longoni, L., Melillo, V., & Papini, M. (2014b). Remote sensing for landslide investigations: An overview on recent achievements and perspectives. *Remote Sensing*, *6*(10), 9600–9652. doi:10.3390/rs6109600.

Teunissen, P. J. G. (2000). *Testing theory: An introduction*. Series on mathematical geodesy and positioning. The Netherlands: Delft University Press.

Tofani, V., Hong, Y., & Singhroy, V. (2014). Introduction: Remote sensing techniques for landslide mapping and monitoring. In K. Sassa et al. (Eds.), *Landslide science for a safer geoenvironment* (Vol. 2, pp. 301–303). Berlin, Heidelberg: Springer.

Van Westen, C. J., Castellanos, E., & Kuriakose, S. K. (2008). Spatial data for landslide susceptibility, hazard, and vulnerability assessment: An overview. *Engineering Geology, 102*, 112–131.

Yin, Y. P., Wang, F. W., & Sun, P. (2009). Landslide hazards triggered by the 2008 Wenchuan earthquake, Sichuan, China. *Landslides, 6*(131), 151.

Zhang, L., Liao, M., Balz, T., Shi, X., & Jiang, Y. (2014). Monitoring landslide activities in the three gorges area with multi-frequency satellite SAR data sets. In M. Scaioni (Ed.), *Modern technologies for landslide investigation and prediction* (pp. 181–208). Berlin, Heidelberg: Springer.

Part I
Ground-Based Monitoring Techniques

Close-Range Photogrammetric Techniques for Deformation Measurement: Applications to Landslides

Marco Scaioni, Tiantian Feng, Ping Lu, Gang Qiao, Xiaohua Tong, Ron Li, Luigi Barazzetti, Mattia Previtali and Riccardo Roncella

Abstract In this chapter, the application of close-range photogrammetry for deformation measurements in the field of landslide investigation and monitoring is discussed. Main advantages of this approach are the non-contact operational capability, the large covered area on the slope to analyze, the high degree of automation, the high acquisition rate, the chance to derive information on the whole surface, not limited to a few control points (area-based deformation measurement), and, generally, a lower cost with respect to 3D scanning technology. Applications are organized into two categories: (1) surface-point tracking (SPT) and (2) comparison of

---

M. Scaioni (✉) · T. Feng · P. Lu · G. Qiao · X. Tong · R. Li
College of Surveying and Geo-Informatics, Tongji University, 1239 Siping Road, Shanghai 200092, People's Republic of China
e-mail: marco.scaioni@polimi.it

T. Feng
e-mail: fengtiantian@tongji.edu.cn

P. Lu
e-mail: luping@tongji.edu.cn

G. Qiao
e-mail: qiaogang@tongji.edu.cn

X. Tong
e-mail: tongxhtj@tongji.edu.cn

R. Li
e-mail: rli@tongji.edu.cn

M. Scaioni · L. Barazzetti · M. Previtali
Department of Architecture, Built Environment and Construction Engineering, Politecnico di Milano, via Ponzio 31, 20133 Milan, Italy
e-mail: luigi.barazzetti@polimi.it

M. Previtali
e-mail: mattia.previtali@mail.polimi.it

R. Roncella
Department of Civil, Environmental, Land Management Engineering and Architecture, University of Parma, via Parco Area delle Scienze, 181/a, 43124 Parma, Italy
e-mail: riccardo.roncella@unipr.it

M. Scaioni (ed.), *Modern Technologies for Landslide Monitoring and Prediction*, Springer Natural Hazards, DOI 10.1007/978-3-662-45931-7_2

surfaces obtained from dense image matching. Different camera configurations and geometric models to transform points from the image space to the object space are also discussed. In the last part of the chapter, a review of the applications reported in the literature and two case studies from the experience of the authors are reported.

Keywords Terrestrial photogrammetry · Computer vision · Deformation measurement · Image metrology · Landslides

1 Introduction: Remote Sensing for Landslide Investigations

Landslides represent a severe threat for the security of people, goods, and properties, as well as for the conservation of the natural and anthropogenic environment (Crudes and Varnes 1996). Monitoring the evolution of an unstable slope generally may provide important information to achieve a better knowledge of the active processes that could lead to failure and to forecast potential geo-disasters (Mantovani et al. 1996). The integration of the observed data into predictive models is the usual approach toward prediction of landslides (see, e.g., Hong et al. 2014), although in many cases the mere observation of the outputs from monitoring sensors may help recognize high-risk conditions.

Several different sensors can be applied to diversified typologies of landslides, see a recently revised classification published by Hungr et al. (2014). Broadly speaking, sensors could be categorized as function of multiple properties, for example, the observed quantity (geometric deformation, geotechnical, hydraulic or geological property, environmental parameter), the location (in situ—on surface or underground—and *remote* sensors from different platforms like ground-based/terrestrial, airborne—onboard of helicopters, aircrafts, UAV's—spaceborne, and terrestrial mobile mapping vehicles), the rapidity of the output delivery (real time, quasi-real time,[1] off-line), and the way outputs are transmitted to the users (manual reading, automatic reading with internal data storage or online communication to a remote control station). Sometimes, long archives of data are needed to obtain up-to-date observations, such as in the case of multi-temporal InSAR processing (see Wasowski and Bovenga (2014)).

[1] Although in the common technical language, the term 'real time' is used for sensors that are able to provide immediate outputs, in landslide monitoring the time frequency should be referred to the speed of the observed process (see Scaioni et al. (2014b)). Consequently, an earth observation system able to measure the deformation of slow-moving landslides with periodical monthly rate measurements can be considered a real-time system. Generally, the term 'quasi-real time' is used for those observations where some tasks of the measurement process introduce a short delay in the output of results. However, in 'quasi-real time' systems such delay does not influence the exploitation of the outcomes.

Any sensors may be installed and controlled independently from the others or may be organized into sensor networks (see Froese and Moreno (2014)).

In particular, *non-contact remote-sensing (RS)* techniques for landslide investigations have become quite popular not only for the purpose of scientific research, but to give an essential contribute to the daily practice of bodies and institutions dealing with the management of geo-hazards (Scaioni 2013). A typical categorization of the state-of-the-art applications into three groups was proposed by Mantovani et al. (1996) and followed in the review published by Metternicht et al. (2005): (1) *landslide recognition, classification, and inventory mapping*; (2) *monitoring of existing landslides*; and (3) *landslide susceptibility mapping and hazard assessment*.

The advantages of RS for *monitoring* applications (see Delacourt et al. 2007) are manifold, including the lack of the necessity to access the slope, unless for installing a few reference points for geo-referencing purpose if required, the coverage of wide areas, the availability of scheduled data acquisition plans (for satellite RS) not requiring specific missions, and the general contained costs, with the exception of those applications that call for long temporal series of high-resolution imagery.

These aspects become even more important when the objective of the monitoring project is the *measurement of deformations*. In such case, the use of *terrestrial remote-sensing techniques* which are operated from ground-based/terrestrial sensors (chiefly digital cameras, terrestrial laser scanners, and ground-based InSAR systems) can be adopted to enlarge both the geometric and temporal resolution of the observations, albeit they require to install a sensor in the proximity of the slope to monitor, sometimes embedded in a permanent, long-term data acquisition system.

Three sensor types are the most involved in terrestrial RS. In order of relevance, the paramount role is taken by terrestrial laser scanning (TLS) (see for example, Bitelli et al. (2004), Guffar et al. (2013), and Barbarella et al. (2014)). A few experiences with laser scanning systems implemented in terrestrial mobile vehicles (mobile laser scanning—MLS) have been also reported in De Agostino et al. (2012) and Tommaselli et al. (2014). The advantage of laser scanning techniques is they may directly provide a dense point cloud describing the topographic ground or vertical rock faces (see Abellán et al. (2014)). This capacity has resulted in a large popularity of such sensors for deriving digital surface models (Pirotti et al. 2013), and for the extraction of geomorphological features, (see Jaboyedoff et al. (2012)). On the other hand, in the case of *deformation measurement*, any kinds of laser scanning approach works well when deformations exceed the measurement accuracy. When small displacements at millimeter size have to be measured from ground-based sensors, like in rock fall assessment, TLS cannot cope effectively in a direct way. Several attempts have been made to exploit the local redundancy of measured points in order to filter out the measurement noise (see, e.g., Scaioni et al. 2013; Abellán et al. 2014), but such applications still have the character of scientific studies focused on specific sites, rather than to be mature for extensive application in the daily practice. In addition, TLS instruments are still quite cumbersome and may be difficult to be operated in harsh mountain environments.

The ground-based interferometric SAR sensors (GBSAR, see Montserrat et al. 2014) coupled with *advanced differential InSAR techniques* (A-DInSAR, see Crosetto et al. 2005) may provide a high accuracy (millimeter level) on the specific site where the instrument is positioned. The instrument may directly yield range displacements along the line of sight (LOS) of those surface points (the so-called *coherent scatterers*, according to Wasowski and Bovenga (2014) which offer good scatter and high persistence along time. Because of the wavelength of the signal in the microwave spectrum, GBSAR can operate at night, with fog, clouds, and with adverse weather conditions. Unfortunately, in the real cases, A-DInSAR suffers from its 'opportunistic' nature (Crosetto et al. 2011), i.e., observed points can be extracted only in locations where a combinations of favorable conditions exist (sensor position, slope orientation, material, and roughness of the ground surface). Furthermore, A-DInSAR needs a long stack of images to model atmospheric errors (see Wasowski and Bovenga (2014)). The main problems of current GBSAR sensors are related to the complexity of the operational systems, which are still difficult to be operated, although the latest instruments seem to become lighter and easier to be carried. Secondly, any instrument has a measurement ambiguity which depends on the signal wavelength and the adopted processing techniques. This problem may become quite relevant when a GBSAR is not used in continuous mode, but it is re-positioned at any observation epochs (Crosetto et al. 2014). If the surface deformations are too fast, some ambiguities may occur in the outcome of GBSAR. To cope with this problem, coupling GBSAR to TLS seems a promising solution, which also helps the visualization and interpretation of point displacements (Wujanz et al. 2013b). Recently, GBSAR measurements have been also exploited in *non-interferometric* mode, even though this application has been done for measuring a few artificial *corner reflectors* installed on the slope (Monserrat et al. 2013). This direction seems to be quite promising because it does not suffer from ambiguity problems typical of microwave RS, although they may provide a lower precision than the interferometric techniques.

The third category of sensors comprehends different kinds of cameras from ground-based platforms which may gather imagery to be used for extracting 2D and 3D displacements on the basis of *image-based deformation measurement* (IBDM) techniques (Scaioni et al. 2014c). The *close-range* (or *terrestrial*) *photogrammetry* (CRPh) is the science dealing with measurement and 3D reconstruction from images (Luhmann et al. 2013). This discipline is strictly related to the research domain of geomatics on one side and on Computer Sciences and Information Technology on the other.

Generally speaking, the use of imagery as data source for measurement may feature several advantages: Images allow storing a huge quantity of information, depending on their resolution; images are easy to understand and interpret, unlike point clouds or single-point observations; images allow off-line measurements; and the acquisition of image does not require the contact with the object, unless for positioning some control points (if strictly required by the adopted procedure).

The focus of this paper is concentrated on RS applications based on the use of close-range photogrammetric techniques for measurement of slope deformations. For this reason, in the next section, a short outlook on the topic is addressed.

1.1 Photogrammetry for Deformation Measurement

Photogrammetry is the science for reconstructing the shape of objects from images. In particular, in close-range photogrammetric applications, digital cameras are stationed on the ground in order to capture images of the target object following a suitable *network geometry* for 2D or 3D reconstruction. The difference with respect to airborne photogrammetry (APh) which is commonly applied to accomplish projects for topographic mapping does not only consist in the targeted objects and the image scale, which usually is much larger in CRPh than in APh because of the shorter acquisition distance. The wide spectrum of applications and object types which can be faced in close range entails the acquisition of more complex network geometries, a much lower standardization of the projects, and in many cases a low available budget. In addition, in modern APh, the use of direct positioning sensors has dramatically simplified the orientation procedures, which are a basic task of any photogrammetric projects. Such sensors are not yet fully mature for the most CRPh applications, except when cameras are implemented in mobile mapping vehicles (Tao and Li 2007). The net subdivision of APh and CRPh with their own specific peculiarities is going to be partially overcome in the future, because of the diffusion of new airborne platforms like *unmanned aerial vehicles* or UAV's (Eisenbeiss and Sauerbier 2012; Colomina and Molina 2014; Nex and Remondino 2014) and airborne oblique images (Dewez 2014; Zhao et al. 2014), which encompass properties from both branches of the photogrammetry.

Disregarding the description of specific methods and algorithms adopted in CRPh, the reconstruction process chiefly consists of the following steps. First of all, cameras generally need to be calibrated to know the internal sensor geometry (inner orientation—IO) and to remove the effect of lens distortion. In CRPh, this is usually carried out beforehand a photogrammetric project by adopting one of the available rigorous photogrammetric procedures for *camera calibration* (Fraser 2013).

The second important stage of the photogrammetric pipeline is the exterior orientation (EO) of the images, which consists in the establishment of a mathematical relationship between the *image space* and the *object space* where the object has to be reconstructed. Typically, in digital images, the former is defined by the imaging sensor geometry, which is made up of a regular grid of sensible elements which output a raster image whose lines and columns can be used as geometric reference. The latter may be fixed by the user on the object space, for example, by independent measurement of a few ground control points (GCP's), or arbitrarily assumed up to a seven parameter similarity transformation. In some projects, the definition of the scale of the reconstructed object may suffice for the expected purpose: In such case, the measurement of a known distance is enough to set up the *datum*. In projects for deformation measurement, the photogrammetric process is repeated (totally or partially, depending on whether the sensors are fixed or not) and metric results obtained at different times (or *epochs*) compared to detect changes. This solution requires the setup of a spatial datum stable over time, where reconstructed coordinates can be referred to.

As outlined in Scaioni et al. (2014c), three main network configurations are commonly used for IBDM projects.

Single-camera (or *monocular*) systems may be used to carry out 2D metric reconstructions in the case of flat or quasi-flat objects (see one example in Scaioni et al. 2014a) or to qualitatively highlight the deformation regions of three-dimensional objects (see Feng et al. 2012). Also in Luhmann (2009), a solution to derive 3D outputs from a single-camera image sequence has been discussed, which was based on the use of special six degree-of-freedom targets. Metric reconstruction may be obtained by using a *homographic* transformation (see Luhmann et al. 2013); otherwise, the processing is completed in the image space and the average ground sample distance (GSD), i.e., the footprint of the pixel size on the ground, adopted for rough scaling. Sometimes, the use of a digital elevation model (DEM) of the object can be used to project in 3D object space the 2D displacement vectors computed from single-camera sequences (see Travelletti et al. (2012)).

In *stereo-camera systems*,[2] a pair of sensors is adopted. After calibration, images may be relatively oriented by simply measuring a set of corresponding points which do not require the knowledge of their coordinates in 3D object space. On the other hand, the knowledge of relative orientation may be used to derive the 3D coordinates of any points (or other features) which can be identified in both images, to be rendered in a three-dimensional arbitrarily located datum. The knowledge of a distance (from example, the camera *baseline*, i.e., the distance between both sensors) can be used to fix only the scale of the reconstruction project, or at least three GCP's are required to totally fix the datum ambiguity. Several models can be implemented for the mathematical modelling of relative orientation, as can be found in the photogrammetric and computer vision literature (see Luhmann et al. 2013; Hartley and Zissermann 2006). Relative orientation may be also seen as a subcase of the general multi-station solution illustrated in the following. Despite of the low redundancy of the photogrammetric network, the technical and budgetary aspects of stereo-camera systems are quite convenient in the practice, especially for installation of permanent image-based monitoring systems where 3D observations are necessary (see Roncella and Forlani 2014). Also stereo-camera systems have been widely used in laboratory experiments (see a review in Scaioni et al. 2013a, 2014b).

Eventually, *multi-station networks* are used when the scene to depict is quite large or when a hyper-redundancy of observations is required (see Fraser et al. (1996)). Rarely, such configuration is implemented along with a fixed multi-camera

[2] Despite of the name, *stereo-camera systems* are not usually aimed at obtaining the stereoscopic vision and cameras may be convergent to improve precision along depth (Fraser 1996).

system (generally limited to three or four sensors), but the same sensor is consecutively placed at different stations (see Previtali et al. (2014)). Of course, a fixed system is needed when the process to monitor cannot be considered stable during the completion of a single-epoch photogrammetric acquisition. Orientation of multiple stations is generally carried out through the *bundle adjustment* (BA), which may provide the EO parameters of all camera stations. If the geometry of the photogrammetric network is sufficiently robust and several images overlap (more than five), camera calibration and orientation may be accomplished at the same time by using *self-calibration* BA. On the other hand, in the experience of the authors, the separation of both stages has to be preferred, in order to set up a specific optimal configuration for the calibration project.

Once all sensors are calibrated and images oriented, the core application focused at deriving useful information for landslide deformation measurements can be undertaken. In the following sections, the products obtainable from CRPh for landslide deformation measurement are outlined, while the reader is addressed to handbooks and to Scaioni et al. (2014a, c) for details about different sensor configurations.

1.2 Applications of CRPh to Landslides

In general, techniques for deformation measurement to be applied to real landslides must be able to operate without any control points on the ground slope or rock cliff to monitor. This is not a general rule, but in practice, the most case studies call for this requirement in the design of the photogrammetric monitoring system. In scaled-down laboratory experiments, the deployment of coded-targets on the surface or the projection/painting of artificial patterns may help the orientation and the measurement process, especially for accurate measurement of point coordinates (Roncella et al. 2004). Close-range photogrammetry based on *coded-targets* is a well assessed methodology for high-precision metrological applications (Luhmann et al. 2013; Scaioni et al. 2014c), which does not require additional presentation. On the other hand, coded-targets may be difficult to be installed in real-scale slopes, and however, target-based techniques may output results concerning a few control points only.

For this reason, in landslide applications, a preference is given to target-less approaches, in order to extend as much as possible the investigated area. Photogrammetry gives the opportunity to broaden the measurement of deformation to a larger region, while other sensors should be preferred to obtain pointwise observations (for example, contact sensors, geodetic, and Global Navigation Satellite System techniques (see Angeli et al. 2000; Fastellini et al. 2011; Pirotti et al. 2014)). A few GCP's may be used only for establishing the datum, possibly outside the area which is directly interested by the deformation process.

Photogrammetry is a technique that enables the use of *area-based deformation measurement* (ABDM) methods. These are important for studying those processes where single points are not enough to completely describe the phenomenon under investigation, or the point signalization is difficult or impossible (e.g., with very small objects). Moreover, also where pointwise measurements are planned to monitor specific points in well-defined locations, ABDM may give an overview on the remainder portions of the object surface. In this way, also deformations in unexpected positions may be detected.

In the following Sect. 2, the main approaches to carry out ABDM using CRPh are presented and discussed. In the successive Sect. 3, some examples from the experience of the authors are reported. Finally, Sect. 4 will draw some conclusions.

2 Methodologies

In this section, the focus is concentrated on target-less close-range photogrammetric techniques that can be used in landslide monitoring projects. These methods may be roughly subdivided into two main categories:

1. *Surface-point Tracking* (SPT) for the reconstruction of surface displacement field in 2D or 3D; and
2. *Comparison of 3D surfaces* to evaluate volumetric changes.

In the following sections, both categories will be illustrated and discussed.

2.1 Surface-point Tracking

In this case, the goal is to reconstruct the 2D or 3D surface displacement field from a single or stereo sequence of images captured from fixed cameras. Also multi-station fixed camera networks may be implemented, but this solution is not usually pursued because of practical and budgetary problems.

The concept is to track a dense field of natural features that can be automatically recognized on the surface of the slope surface. If the texture is well contrasted, a dense field of observed points may be detected and tracked along different epochs. *Digital Image Correlation* (DIC) or *image matching* techniques (Le Moigne et al. 2011; Baker et al. 2011; Grün 2012) are applied to this purpose, after the completion of preliminary tasks described in Sect. 1.1, i.e., camera calibration and orientation. Some aspects related to this category of applications are described in the following sections.

Corresponding points tracked using SPT techniques can be used to derive *velocity maps*, *direction maps*, and *differential maps*.

2.1.1 Image Matching

The basic concept of different image matching/DIC algorithms is to start from one reference image (called 'master') where a set of 'seed' points is defined by using specific *interest operators* (Jazayeri and Fraser 2010) or simply by considering nodes of a regular grid in the image (or object) space. Corresponding points are sought on a second image, referred to as 'slave.' In SPT application, the same points have to be tracked along multiple images of a sequence. Consequently, the matching process is repeated in any couple of subsequent images. Any 'slave' image, after matching with the previous 'master' is accomplished, becomes at its turn the new 'master' and so on up to the end of the sequence. The 'tracked' points from the initial 'slave' can be used as 'seed' points for next matching. On the other hand, if the density of 'tracked' points is low, new 'seed' points may be instantiated not to lower the density of the displacement field.

Two main categories of image matching/DIC techniques are commonly used: area-based matching (ABM) and feature-based matching (FBM), see Grün (2012). Generally, matching algorithms used in FPT work simultaneously on two images, while in surface reconstruction also *multi-photo* matching is applied (see Sect. 2.2).

Area-based Matching

These algorithms evaluate the similarity by minimizing a cost function based on the intensity values in two windows extracted around one 'seed' point on the 'master' and its predicted approximate position on the 'slave.' Such prediction may depend on the expected displacements. Additional 'geometric' constraints may be also included to increase the robustness. In general, ABM methods may yield a subpixel precision (up to 0.05–0.2 pixel size), but they cannot cope with images showing large perspective deformations (e.g., owing to convergent camera stations) or radiometric changes.

The most popular ABM algorithms are *Normalized Cross-correlation* (NCC) which may be successfully applied when geometric deformations between 'master' and 'slave' images almost consist in shifts (see Rosenfeld and Kak (1976)). The 'pull-in' range capability is quite large, because homologous can be looked for inside a search window which may span over a wide areas on the 'slave.' On the other hand, the larger is the search window size, the higher is the risk of false matching detection.

When other deformations than shifts are significant, *Least-squares Matching* (LMS) technique can be effectively applied (see Grün (1985)). Here, the minimization process incorporates the estimate of a geometric transformation mapping a window (*template*) extracted around the 'seed point' on the 'master' image to fit with the intensity values of a search window located around the predicted position on the 'slave.' Generally, an affine transformation is implemented, but also more involved implementations can be found in the literature (Baltsavias 1991; Bethmann and Luhmann 2010; Roncella et al. 2012). Estimate of linear radiometric transformation can be also included into the Least-squares model, or the preliminary *image equalization* can be applied, for example, by using Wallis filter (Wallis 1976) or *polarimetric filters* (see Guidi et al. (2014)).

Feature-based Matching

These algorithms rely on the application of *interest operators* (Jazayeri and Fraser 2010) to both images in order to extract two sets of candidate homologous points. Matching is based on a similarity criterion which is computed in the nearby of the candidate points. The most commonly used algorithms adopted in FBM are SIFT (Lowe 2004) and SURF (Bay et al. 2008), but an exhaustive review and comparison can be found in Apollonio et al. (2014). Both incorporate the evaluation of a similarity measure (*descriptor*) to be used for matching. More details can be found in Barazzetti et al. (2010).

FBM is usually more competitive than ABM to look for correspondences in images with large geometric and radiometric difference, for example, when convergent images are used. On the other hand, they feature a lower precision with respect to ABM algorithms, in general between 0.3 and 1 pixel size.

In addition to *robustness* with respect to changes in the images, the *repeatability* is another important property when the same point has to be tracked in a long image sequence. For example, *Förstner operator* (Förstner and Gülch 1987) suffers from large deformations in scaling and rotation, but it is more suitable to track the same points in several images (Barazzetti et al. 2010).

Selection of ABM/FBM

The selection of the matching method depends on the properties of the target object and on the velocity of the image content change. If a slow process is investigated with quite regular changes between consecutive images, ABM has to be preferred. If sudden changes may happen, FBM is generally more appropriate.

2.1.2 Camera Configurations in SPF

In the case a single sequence is used, point matching has to be only accomplished between *asynchronous* images with the aim to detect and measure the displacements over time. The *homography* model can be exploited to derive 2D coordinates in object space if the object surface is flat or quasi-flat (see Scaioni et al. (2014a)). In the case of non-flat surfaces, SPT may only give a qualitative output but not any metric information.

When two (or more) cameras are contemporarily used, corresponding points have also to be found across *synchronous* images taken from different standpoints. This task is generally more complex because the perspective and radiometric

deformations between synchronous images have to be kept into consideration. Thanks to the knowledge of relative orientation between the images, 3D coordinates of points can be then derived. In addition, the *synchronization* of different cameras has to be carefully considered when dealing with stereo or multi-station networks, and the surface process to be imaged is fast. This is not a real issue in real-field applications, because the synchronization delay usually accounts for a fraction of second, time during which the slope may be considered as completely stable. On the other hand, this may become a serious issue when dealing with laboratory experiments where a high-dynamic failure process is triggered. Solutions to cope with this problem are addressed in Spencer and Shah (2004) and Raguse and Heipke (2009).

2.2 Comparison of Digital Surface Models

Dense image matching algorithms can be exploited to derive a digital surface model (DSM) that, if characterized by proper accuracy and resolution, can be considered as a faithful numerical description of the slope topography (see Haala (2013) and Remondino et al. (2014)). Starting from at least two images with known EO, image matching techniques are applied to find corresponding points, whose spatial intersection gives 3D coordinates in object space. The resulting set of 3D points constitutes the so-called *point cloud*, which can be interpolated to derive a DSM. A brief review of dense image matching process is reported in Sect. 2.2.1.

The knowledge of two DSM's of the same slope gathered at different times may be used for detecting changes and deformations such as those related to slope failures. Unfortunately, the comparison of two DSM's cannot be operated by simply comparing coordinates of corresponding points. This is due to the fact that the 'seed' points used for matching (see Sect. 2.1.1) may be different between the epochs to compare, because of the changed texture, illumination, vegetation, and the like. In addition, very often the point cloud directly obtained from dense matching is interpolated in order to fill gaps and/or to decimate points in areas where they are too dense. As a result, the comparison of DSM's needs the application of ABDM techniques, as described in Sect. 2.2.2.

2.2.1 Dense Image Matching

Several dense matching algorithms have been proposed, as reviewed in Scharstein and Szeliski (2002), Seitz et al. (2006), Grün (2012), Haala (2013), and Remondino et al. (2014).

The first key point of different methods relies in the way they provide a dense number of points to guarantee a continuous reconstruction of the surface also in presence of sharp edges or low-texture areas. The second key point is related to the robustness against outliers, which may be due to a long list of causative reasons: repeated patterns, poor contrast in the images, change of illumination, large

perspective deformations, moving objects, image blur, and noise. Both properties have to be possibly fulfilled. For example, FBM techniques are usually not suitable for dense matching because they are robust, but they cannot cope effectively with the required high point density. It deserves to be mentioned that the network geometry also plays an important role to achieve both key points. In presence of a multi-station network, the better coverage of the object and the higher redundancy may help in both directions. But if a stereo-camera system is the only solution owing to technical or budgetary motivations, any algorithms will face more difficult operating conditions.

The point density can be achieved by using *hierarchical matching strategies*, by integrating different matching strategies (for example, ABM, FBM, and line matching as proposed in Zhang 2005) or *pixelwise matching* (like Semi-global Matching—SGM, since Hirschmüller 2005).

The main approaches to cope with outliers are based on redundant images which do overlap the same portion of the object to reconstruct. Examples are Multi-photo Geometrically Constrained Matching (MGCM) of Grün and Baltsavias (1988), Geometrically Constrained Cross-correlation (GC^3) of Li and Grün (2004), and PMVS (Furukawa and Ponce 2010). Also SGM has been implemented in *multi-view* fashion (see a review in Bethmann and Luhmann (2014) and Dall'Asta and Roncella (2014)).

Any dense matching techniques are based on the minimization of a *cost function* that is built up on the basis of the intensity values. Many methods also include some constraints coming from the known image mutual geometry (i.e., the EO) (see Maas (1996)).

The precision that can be obtained is difficult to be predicted and to be evaluated. This depends on the large number of factors that may have influence on it, first of all the quality and the texture of the images (see Grün (2012)). Some attempts have been accomplished to compare different methods, but results obtained so far cannot cover all existing techniques and possible object configurations (see Seitz et al. (2006), Remondino et al. (2014), and Toschi et al. (2014)).

2.2.2 Area-Based Deformation Measurement

Point clouds and DSM's are usually compared in a pairwise manner. This means that in the case of a sequence made up of n several reconstructed scenes (S_k), the comparison may be carried out with respect to a reference epoch ('many-to-one') or using a *sequential* method.

As already introduced at the beginning of Sect. 2.2, the point clouds obtained from image-based dense matching cannot be compared point-by-point (see Lindenbergh (2010)). On the other hand, this problem also happens when comparing point clouds obtained from laser scanning or 3D imaging sensors (Pears et al. 2012; Remondino and Stoppa 2013; and Araiba and Sakai 2014), as discussed in Scaioni et al. (2013a).

In general, two point clouds coming from the photogrammetric reconstruction of *ground slopes* may feature changes due to global or local sliding, to surface erosion or excavation, and to debris accumulation and vegetation growth. In addition, the presence of rigid-body transformation may be related to registration errors. This is due to the fact that landslides on soil generally do not move as a whole, but the surface may show different non-rigid shape deformation. In the case of *rock slopes* and *rock cliffs*, the entire rock bulk could also move as a unique body, and consequently, the discrimination of the different deformation components is more involved. Furthermore, in such cases, the surface displacements may be quite small (millimeters), calling for the use of specific techniques (see Wu et al. (2012), Scaioni et al. (2013a), and Abellán et al. (2014)).

Given two DSM's (S_i and S_j) to be compared, one of the following approaches may be pursued, if they have been already geo-referenced into the same reference frame, as it happens when the EO of photogrammetric blocks has been already established with respect to an external *datum*:

1. *Point-to-point* analysis: For each point in S_i, the closest point in S_j is selected, and then, the distance between both points is computed;
2. *Point-to-surface* analysis: The point cloud assumed as reference (S_i) is interpolated or approximated to derive a continuous mathematical surface, e.g., using triangulated irregular network (TIN) mesh, NURBS, 2D splines, and other parametric functions (see Hoffmann (1989)). Then, per each point in point cloud S_j, the *minimum distance* with respect to the surface of S_i is computed;
3. *Surface-to-surface* analysis: Both point clouds S_i and S_j are interpolated or approximated to yield two mathematical surfaces that usually are referred to a reference plane π, e.g., the horizontal plane or the average plane of the slope/cliff. Consequently, each point cloud can be now represented using a DEM, where the DoF of any point are the distance z with respect to plane π and the position x–y in the plane. Changes can be detected from the computation of the volume comprised between both DEM's (method '3a' (see Xue et al. 2014)) or by computing the so-called ΔDEM, which is the difference between DEM's derived at both epochs S_i and S_j (method '3b') (see Gu et al. (2014)); and
4. *Fitting of local regular surfaces*: In the case the noise is larger with respect to the magnitude of sought displacements, a solution is to locally fit a set of regular surfaces (usually planes) to the raw point clouds S_i and S_j; and the comparison is then accomplished through the statistical analysis of the surface parameters (see Lindenbergh and Pfeifer (2005) and Scaioni et al. (2013c)).

The selection of one model depends on many conditions, and it is difficult to find a ubiquitous best solution. Factors that should be considered are as follows:

- *Direction of displacements*: method (2) may provide displacements only in the local normal direction of the reference surface; methods in group (3) can provide displacement only in the local z direction with respect to the reference plane, while it cannot give information of pure shifts along x–y;
- *Noise*: Methods (1) and (2) are more sensitive to noise in the point cloud;

- *Point density*: All methods work better with high point density, but method (1) suffers more from a low density;
- *Surface roughness*: Similarly to the dependency on noise, methods (1) and (2) are more sensitive to rough surfaces;
- *Sharp breaklines*: Methods based on interpolation suffer from the presence of breaklines, especially in the case of (even small) misalignments; and
- *Gaps or holes in the data sets*: If the surface is regular, methods (3) may mitigate such effect, while others will not provide any outputs in corresponding regions.

In the case geo-referencing of both point clouds is pending or not coherent, a preliminary relative registration has to be carried out, for example, by using ICP algorithm (see Pomerleau et al. (2013)). Because potential changes may affect both point clouds S_i and S_j, the registration algorithm should be applied only in the case some stable regions may be identified. Robust techniques that were developed for registration of point clouds when changes had occurred are also available (see Wujanz et al. (2013a)). If the camera stations are fixed and stable enough not to change their relative and absolute position with respect to the object, the use of GCP's can be avoided.

The *statistical analysis* of computed deformations is an important aspect not to be forgotten. While a consolidated background exists in the case of regressions and geodetic networks (Teunissen 2000), in the case of surface deformation the main contribution is due to Dermanis (2011).

It is also worth to mention that two or more point clouds may be also compared in alternative ways. For example, Ghuffar et al. (2013) proposed to track *3D range flow* to analyze the spatial displacement field of a landslide. Other authors (Barbarella et al. 2014) proposed the comparison of cross sections for reducing the spatial dimensionality of this problem.

3 Some Examples

In this section, some experimental examples of techniques described in the methodological Sect. 2 are reported. First of all, a brief review on the applications that can be found in the literature is presented in Sect. 3.1.

3.1 Review of the Literature

Photogrammetry has found several applications for the analysis of ground deformations over landslides (see Delacourt (2007)). Although this chapter is focusing on close-range applications based on ground-based cameras, it is worthy to mention a few significant achievements obtained from aerial or satellite imagery. Some

methods implemented to extract information from data gathered by these platforms, indeed, are quite similar to the ones described in the previous section of this chapter in the case of terrestrial images.

Aerial photogrammetry has found application for the reconstruction of digital terrain models (DTM's) to be compared over time (Casson et al. 2005) or to produce *orthoimages* useful to track horizontal displacements (see Wiegand et al. (2013)). The diffusion of UAV platforms found an increasing popularity in recent years thanks to the capability of flying at low height to enlarge the photo scale (see Niethammer et al. (2012)).

Satellite high-resolution and *very high-resolution optical imagery* have been used for tracking surface displacements (LePrince et al. 2008), also in combination with aerial images (Delacourt et al. 2004; Debella-Gilo and Kääb 2012). The output mainly consists in the analysis of surface displacement field or cover change (for example, for the detection of shallow landslides). The trend is to use ABM (see Rosu et al. (2014)). The comparison of multi-temporal DTM's has not been exploited yet due to the unsufficient accuracy. On the other hand, a DTM may assist to remove the topographic components of surface displacements obtained from the satellite high-resolution imagery, as in the case of Stumpf et al. (2014).

In the field of *close-range photogrammetry*, a few fixed camera systems are described in the literature. *Single-camera* systems are reported in Ohnishi et al. (2006), who deals with the design of the photogrammetric system, Delacourt et al. (2007), Travelletti et al. (2012), and Motta et al. (2013). In particular, Travelletti et al. (2012) describe a system installed in front of the Super-Sauze landslide (South French Alps) and provide a thorough analysis of such kind of application. *Stereo-camera* systems are reported in Roncella and Forlani (2014) for monitoring of a real-scale slope, while in Roncella et al. (2004) and Feng et al. (2012) a couple of implementations for observation of simulated landslides in a laboratory facility are illustrated. Some results obtained from this system are outlined in Sect. 3.2. Applications of *multi-station networks* based on non-fixed cameras are reported in Cardenal et al. (2008) and Previtali et al. (2014). In both cases, photogrammetry is used for reconstruction of multiple DTM's to be compared over time and detect volumetric changes. The latter case is presented in Sect. 3.3. In Akca (2013), a photogrammetric network composed of four high-speed industrial cameras was implemented for surveying a simulated shallow landslide. As a final comment on the use of close-range photogrammetric techniques for landslide deformation measurement, existing systems are based on the analysis of volume changes (stereo-systems) or on SPT (single-camera systems). The integration of both 2D and 3D outputs, which are complementary, is expected to enrich the total information available. In addition, a wider use of robust FBM techniques may increase the capability of SPT, as proposed in Feng et al. (2012).

Fig. 1 The landslide simulation platform established at Tongji University (Shanghai, P.R. China)

3.2 Application in a Scaled-down Landslide Simulation Platform

The first example is related to the analysis of image sequences gathered by stereo-camera systems implemented in a landslide simulation facility established at Tongji University (Shanghai, P.R. China) (see Scaioni et al. (2013b) and Qiao et al. (2013)). Aim of these experiments was to set up a spatial sensor network to be included in early-warning system (see, e.g., Intrieri et al. 2012) for a real landslide. Here, we focus on the photogrammetric subsystem, consisting in a couple of stereo-camera systems, one based on a pair of Nikon D200 single-lens-reflex (SLR) cameras (CCD sensor 2,896 × 1,944[3] pixels, 8.1 μm pixel size, 35-mm focal lens) and the other based on a pair of DALSA Falcon 4M60 industrial monochromatic cameras (CMOS sensor 2,352 × 1,728 pixels, 7.4 μm pixel size, 24-mm focal lens). The former system was used for capturing images all along the experiment run at low acquisition rate (0.033 Hz), and the latter was focused on capturing the last part of the experiment at high rate (20 Hz) during the final collapse of the whole scaled-down slope. Two images of the simulation platform and the implemented stereo-camera systems are shown in Figs. 1 and 2, respectively.

While a set of contact geotechnical and deformation sensors where buried in the slope to record pointwise observations, both camera systems were aimed at the description of 'area-based' processes. In particular, the recognition of useful information for predicting failures was the main target.

SPT was used to highlight the regions where the sliding process started. The main principle was based on the assumption that, in good conditions, the stable areas of the slope should be characterized by a huge number of corresponding points extracted by means of FBM. Preference was given to this category of matching techniques due to their major robustness, also considering the transfer of this system to monitor a real-scale landslide. On the other hand, stable

[3] The full resolution of the sensor (3,872 × 2,592 pixel) was not used.

Fig. 2 The stereo-camera systems implemented in front of the *scaled-down slope* of the landslide simulation platform

corresponding points in stable areas are characterized by non-significantly changed image coordinates in consecutive images, while their density and spatial distribution may outline the extension of such areas. When surface displacements start they are expected to be smaller and then to increase afterwards. Consequently, in the initial stage of a small or large slope failure, FBM could extract several corresponding points between asynchronous images in slowly moving areas, but this time featuring small displacements that significantly differ from zero. The knowledge of the time tag of any image of the sequence can be exploited for computing the velocity of any points. Velocity maps and/or statistics on moving points could be used as indicators of forthcoming failures. When a failure happens, also FBM is not able to detect corresponding points.

FBM has been implemented by using SURF (Bay et al. 2008), which may provide slightly less interest points than SIFT, but with a smaller computational cost. Although two stereo-systems were used, SPT was independently carried out in single sequences. This solution prevented the extraction of precise metric 3D information, because the surface of the scaled-down slope was not flat and the *homography* model could not be applied for image rectification. The availability of the synchronous images gathered by a second camera could be used for computing 3D coordinates of points that, however, were retained a redundant information in this application focused at detecting the areas where surface displacements were accelerating. In addition, stereo-matching also requires looking for corresponding points between synchronous image pairs, involving an additional computational burden and the risk of reducing the number of extracted points.

In Fig. 3, the methodology adopted for the interpretation of SPT is illustrated. Figure 3a shows an image captured by the 'left' camera Nikon D200. Figure 3b represents the successive scene, captured 30 s after. During this time, the scaled-down slope was completely stable, so that the SURF-based matching provided a

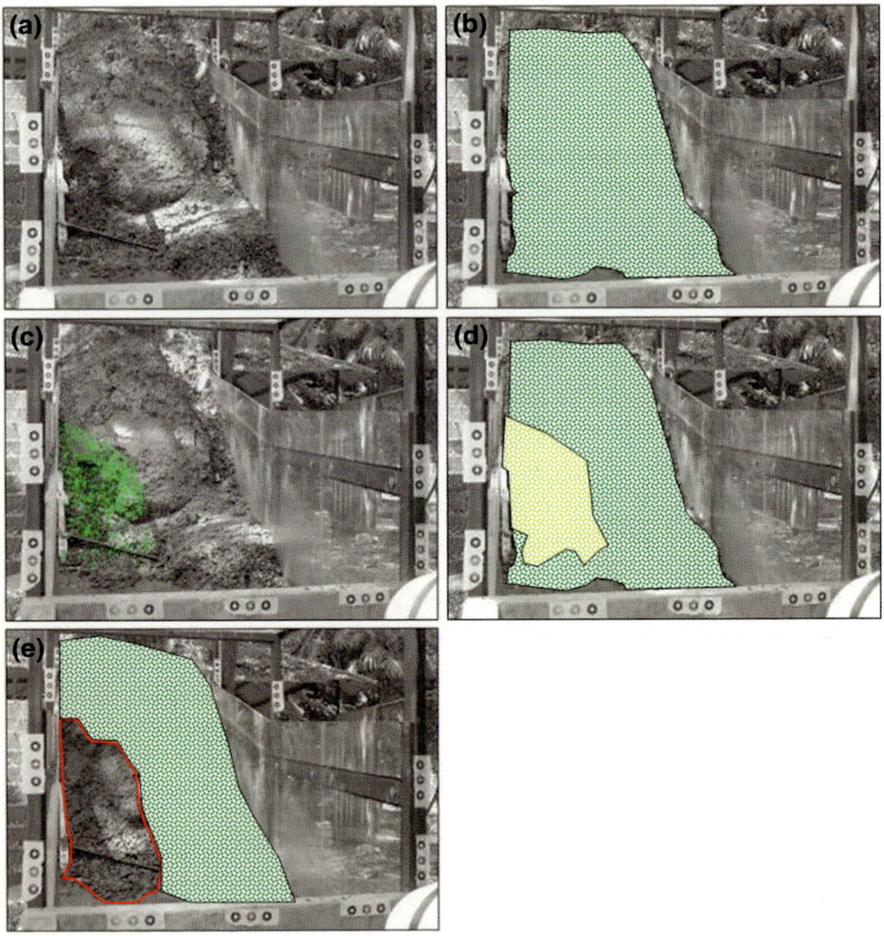

Fig. 3 Example of the interpretation of SPT results on an image sequence capturing a *scaled-down* landslide simulation experiment at Tongji University, Shanghai (P.R. China)

dense filed of extracted points, whose coordinates did not significantly differ from the ones of corresponding points in the previous image (green area in Fig. 3b).

Figure 3c shows an area where a local displacement started, as highlighted by the green points which have been extracted using FBM with respect to the previous image of the same sequence. These points were characterized by significant displacements, unlike points extracted on the remainder portion of the slope (not drawn here) which were still stable. The interpretation of this output is shown in Fig. 3d. The green area is stable, while the yellow area is moving. After a few scenes, the yellow area collapsed and no surface points could be tracked in it (red area in Fig. 3e). This failure involved also some close areas which had kept stable before this collapse.

This processing could be used for understanding the ongoing failures and to predict them on the basis of the number and the speed of points. Moreover, the information derived from SPT could be integrated with the one achievable by a sensor network of geotechnical instruments (Scaioni et al. 2013b) and by other photogrammetry outputs such as DSM and lateral profiles (Scaioni et al. 2014a).

3.3 Monitoring of a Slope Using Multi-station Photogrammetry

In the case, the area to monitor over time is large and a multi-station network is required. In the second example reported in this section, the purpose was to study the debris movement in the Tartano Valley (Italian Alps). Here, the land erosion on the bed of Tartano River, on the flanks of the main valley, and the contribution from the tributary creeks may result in debris accumulation. This increases the risk of important debris flow events whose runoff could reach the inhabited areas in the Adda River Valley, where Tartano River flows.

While the measurement of debris accumulation over a large area is a typical application for airborne laser scanning (ALS—see Heritage and Large 2009), here the specific aim was to investigate a catchment area on a slope sizing 240 m long and 130 m wide (see Fig. 4). The reconstruction of the slope surface at different epochs was needed to highlight the average and localized erosion process and to find a possible relationship with rainfall rate. TLS and digital close-range photogrammetry were the considered solutions to derive such multi-temporal DSM's to be compared. TLS would be the simplest solution to this problem, as proved in many studies (see, e.g., Barbarella et al. (2014) and Mazzanti et al. (2014)). Due to logistic reasons, this option could not be implemented.

A multi-station photogrammetric network consisting of 13 camera poses was then designed with the purpose of reconstructing the three-dimensional surface of the slope using dense matching (see Sect. 2.2.1). A set of GCP's were installed in some accessible positions on the slope, and their position measured with a theodolite during any measurements campaigns (totally three). A calibrated SLR camera

Fig. 4 The slope in Tartano Valley (Italian Alps)

Nikon D700 (CMOS sensor, 4,256 × 2,832 pixels, 8.4 μm pixel size, 90-mm focal lens) was used for image acquisition. The EO of the block of images corresponding to any epochs was computed through a BA incorporating both GCP's and a set of tie points extracted by the automatic procedure ATiPE (see Barazzetti et al. (2010)). In Fig. 5, the camera stations and the extracted tie points are depicted. The density of tie points was also exploited to generate an initial rough DSM useful to instantiate the following dense matching phase. Thanks to the availability of GCP's and the stable geometry of the network, a good repeatability could be obtained even though cameras were not in fixed positions. As reported in Previtali et al. (2014), the analysis of 3D point accuracy after BA always yielded root-mean-square errors (RMSE) lower than 1 cm.

Dense matching was carried out using MGCM+ algorithm of Previtali et al. (2011), which had been developed from the MGCM of Grün and Baltsavias (1988). Of course, this method could be viable because of the redundant number of images with multiple overlaps. This solution helps mitigate the risk of false matchings,

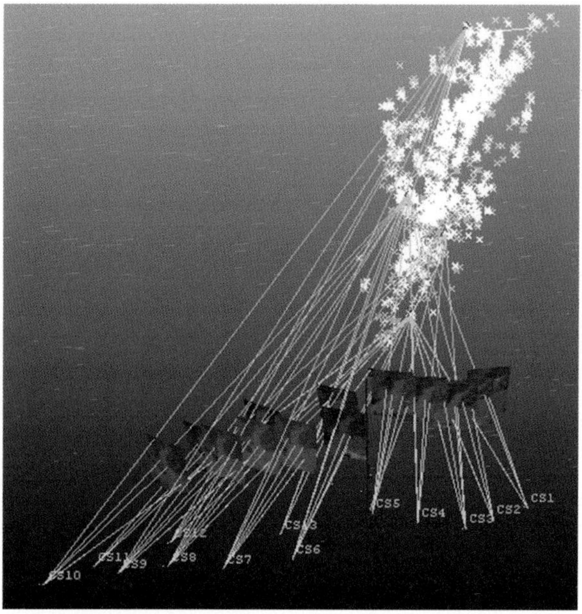

Fig. 5 Typical scheme of the multi-station photogrammetric network adopted at different epochs. Camera stations are indicated with 'CS', tie points adopted for bundle adjustment with white crosses, and GCP's with *red triangles*

which is quite high in the case under discussion because of the presence of repetitive patterns due to soil, rocks, and vegetation.

Surface reconstruction was operated along with an iterative approach. This was necessary because MGCM+ required an approximate surface to locate the rough position of corresponding points in multiple overlap images. Some criteria were implemented to select the best images to be used per each portion of the slope, according to its spatial aspect. The approximate surface is used to compute a TIN, then any faces of this is reprojected on all the images (the EO is known). From the analysis of the shape and the size of any projected triangles and by considering the angles between the normal to the TIN's face and the camera axis, all images were ranked. Images with a too large angle with respect to the surface normal (represented by the normal to the TIN's face) do not positively contribute to the matching process. Furthermore, when more than four overlapping images are available for the same facet, their added value does not compensate for the increase computational burden. We limited here the number of photos to process in the MGCM+ to four, selected on the basis of the computed ranking.

The point density of the required DSM is defined in the object space. In a first iteration, the TIN obtained from the set of tie points (Surface S_{0t}) was used to derive a DSM with a point interdistance of 15 cm × 15 cm (Surface S_{1t}) with respect to the average slope surface. In a second iteration, surface S_{1t} was used as input to be refined up to a final point density of about 2 cm × 2 cm (Surface S_{2t}).

DSM's reconstructed at different epochs (S_{2t}, $t = 1, 2, 3$) were compared between them. The methodology applied for this analysis was based on a *surface-to-surface*

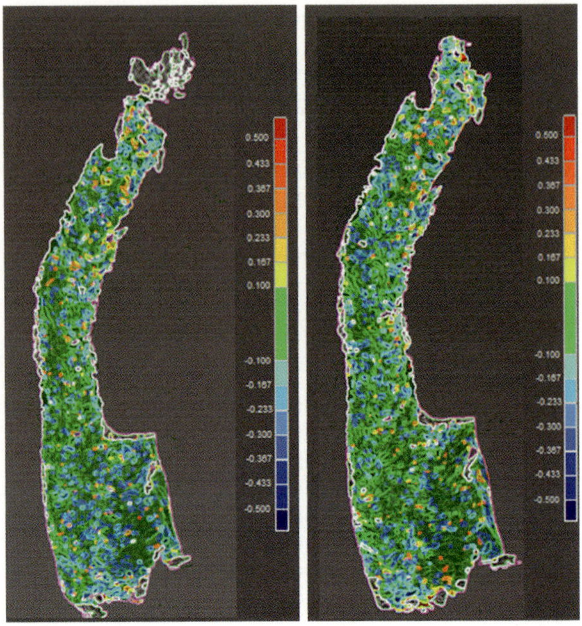

Fig. 6 Two deformation maps corresponding to epoch 2 (August 2010) versus epoch 1 (July 2010) and to epoch 3 (July 2011) versus epoch 2 (August 2010)

approach, motivated by the fact that surface changes could happen in different directions due to erosion and rock movements. A TIN structure was computed by selecting a minimum size of any triangles in order to avoid to model small local perturbances in the digital model. In a second stage, TIN models were compared.

In Fig. 6, the results of surface comparison between S_{21} versus S_{22} and S_{22} versus S_{23} are shown. An estimated precision of about ±15 cm was obtained for the deformation maps. The analysis of the first two DSM's (S_{21} on July 2010 and S_{22} on August 2010) related to the same summer in 2010 does not show significant changes with respect to the estimated uncertainty. When looking at the DSM's corresponding to epochs in two different years (S_{22} on August 2010 and S_{22} on August 2011), major departures are clearly evident. Small areas featuring a negative change from −15 to −25 cm are visible, confirming the ongoing erosion process. The comparison against images confirmed the results of surface-to-surface analysis (see Previtali et al. (2014)).

4 Conclusions

This chapter described the use of image-based techniques typical of close-range photogrammetry for deformation measurement in the field of landslide analysis and monitoring. Methods have been collected in two major categories: SPT and *comparison of surfaces* from dense matching. The former techniques are useful for the reconstruction of the surface displacement field of a slope, in order to take apart among the stable and unstable areas and possibly to timely predict forthcoming failures. The latter is based on the reconstruction of numerical models (digital surface model—DSM) of ground slopes or rock face surfaces at different epochs and their successive comparison and analysis (area-based deformation measurement—ABDM). This second category is more suitable to provide volumetric and three-dimensional changes and it is sensitive to any change of the shape. On the contrary, SPT is sensitive to the change of texture, for example, due to the earth or soil flow in the direction parallel to the image plane. This means that SPT and surface comparison are complementary techniques that in some cases should be contemporarily implemented, especially when strict hypotheses on the nature of the slope failure mechanism cannot be made. Alternatively, SPT could be implemented with 3D scanning sensors, which also can provide the reconstruction of the topographic surface but cannot give detailed information on texture.

Close-range photogrammetry can be applied to both the analysis of real-scale slopes where natural landslides are active (see the example in Sect. 3.3) and scaled-down models adopted for laboratory simulation of landslides (Sect. 3.2).

Three main camera configurations have been analyzed: *single-camera* systems, *stereo-camera* systems, and *multi-station networks*. Only the last two configurations may provide 3D coordinates, unless special cases (see Sect. 1.1). Any of them may be implemented in fixed camera systems, where the registration/orientation can be computed once and then applied to any data sets. This solution is required to measure fast processes or in the case of permanent, long-term monitoring systems. In the case of non-fixed systems, attention should be paid in registering any multi-temporal data sets into the same stable reference frame.

SPT is chiefly applied along with single-camera systems, with potential extension to stereo-camera systems. On the other hand, in such a case, the number of extracted features may be lower due to the necessity of matching feature points along the time and between synchronous image pairs.

Surface comparison based on dense matching requires at least a stereo configuration, but it may take advantage from multi-camera systems owing to the higher data redundancy.

If on one side the application of image-based techniques and in general of the ABDM concepts to landslide investigation will benefit from the technical progress in photogrammetry, computer vision and image processing and understanding, there are a few specific research directions which would call for concentrating future efforts. These entail first the integration between different sensors and methods, not only limited to the image-based solutions, but including 3D scanning and ranging, geotechnical and environmental sensors, and radar interferometry. The organization of different instruments into *spatial sensor networks* is expected to provide complementary, reliable, and cross-validated measurements able to allow a better interpretation of phenomena.

Secondly, the integration between observations and mechanical modelling of slope failure is a key point toward a timely prediction of landslides and the estimate of their magnitude. In addition, such integration is useful either to improve the calibration of numerical models and to check the quality of the observations (*data assimilation*).

Third, the *statistical analysis* of deformation measurement results is an aspect that deserves more attention in the future, being the qualitative analysis the main method used in the applications accomplished so far.

Acknowledgments This research was partially funded by the 863 National High-tech R&D Program of China (No. 2012AA121302) and by the 973 National Basic Research Program of China (No. 2013CB733204). Also, this research was supported by the Italian Ministry of University and Research within the project FIRB—Futuro in Ricerca 2010 (No. RBFR10NM3Z).

References

Abellán, A., Oppikofer, T., Jaboyedoff, M., Rosser, N. J., Lim, M., & Lato, M. J. (2014). Terrestrial laser scanning of rock slope instabilities. *Earth Surface Processes and Landforms, 39*, 80–97.

Akca, D. (2013). Photogrammetric monitoring of an artificially generated shallow landslide. *The Photogrammetric Record, 28*(142), 178–195.

Angeli, M., Pasuto, A., & Silvano, S. (2000). A critical review of landslide monitoring experiences. *Engineering Geology, 55*, 133–147.

Apollonio, F. I., Ballabeni, A., Gaiani, M., & Remondino, F. (2014). Evaluation of feature-based methods for automated network orientation. *International Archives of the Photogrammetry, Remote Sensing and Spatial Information Sciences, 40*(5), 47–54.

Araiba, K., & Sakai, N. (2014). Laser scanner application in monitoring short-term slope deformation. In K. Sassa, et al. (Eds.), *Landslide science for a safer geoenvironment* (Vol. 2, pp. 5–11). Berlin, Heidelberg: Springer.

Baker, S., Scharstein, D., Lewis, J. P., Roth, S., Black, M. J., & Szeliski, R. (2011). A database and evaluation methodology for optical flow. *International Journal of Computer Vision, 92*(1), 1–31.

Baltsavias, E. P. (1991). Multiphoto geometrically constrained matching. *Ph.D dissertation, Mitteilungen Nr. 49*, p. 221. Institute of Geodesy and Photogrammetry, ETH Zurich.

Barazzetti, L., Remondino, F., & Scaioni, M. (2010). Orientation and 3D modelling from markerless terrestrial images: combining accuracy with automation. *The Photogrammetric Record, 25*, 356–381.

Barbarella, M., Fiani, M., & Lugli, A. (2014). Multi-temporal terrestrial laser scanning survey of a landslide. In M. Scaioni (Ed.), *Modern technologies for landslide investigation and prediction* (pp. 89–121). Berlin, Heidelberg: Springer.

Bay, H., Ess, A., Tuytelaars, T., & Van Gool, L. (2008). Speeded-up robust features (SURF). *Computer Vision and Image Understanding, 110*(3), 346–359.

Bethmann, F., & Luhmann, T. (2010). Least-squares matching with advanced geometric transformation models. *International Archives of the Photogrammetry, Remote Sensing and Spatial Information Sciences, 38*(5), 86–91.

Bethmann, F., & Luhmann, T. (2014). Object-based multi-image semi-global matching—concept and first results. *International Archives of the Photogrammetry, Remote Sensing and Spatial Information Sciences, 40*(5), 93–100.

Bitelli, G., Dubbini, M., & Zanutta, A. (2004). Terrestrial laser scanning and digital photogrammetry techniques to monitor landslide bodies. *International Archives of the Photogrammetry, Remote Sensing and Spatial Information Sciences, 38*(7B), 246–251.

Cardenal, J., Mata, E., Perez-Garcia, J., Delgado, J., Andez, M., Gonzales, A., & Diaz-de-Teran, J. (2008). Close range digital photogrammetry techniques applied to landslide monitoring.

International Archives of the Photogrammetry, Remote Sensing and Spatial Information Sciences, 37(B8), 235–240.

Casson, B., Delacourt, C., & Allemand, P. (2005). Contribution of multi-temporal sensing images to characterize landslide slip surface—application to the La Clapiere Landslide (France). *Natural Hazards and Earth System Sciences, 5*(3), 425–437.

Colomina, I., & Molina, P. (2014). Unmanned aerial systems for photogrammetry and remote sensing: A review. *ISPRS Journal of Photogrammetry and Remote Sensing, 92*, 79–97.

Crosetto, M., Crippa, B., Biescas, E., Monserrat, O., Agudo, M., & Fernández, P. (2005). State-of-the-art of land deformation monitoring using SAR interferometry. *Photogrammetrie, Fernerkundung, Geoinformation, 6*, 497–510.

Crosetto, M., Monserrat, O., Cuevas, M., & Crippa, B. (2011). Spaceborne differential SAR interferometry: Data analysis tools for deformation measurement. *Remote Sensing, 4*, 305–318.

Crosetto, M., Monserrat, O., Luzi, G., Cuevas-Gonzáles, M., & Devanthéry, N. (2014). Discontinuous GBSAR deformation monitoring. *ISPRS Journal of Photogrammetry and Remote Sensing, 93*, 136–141.

Cruden, D. M., & Varnes, D. J. (1996). Landslides types and processes. In A.K. Turner & R.L. Schuster (Eds.), *Landslides: Investigation and mitigation* (pp. 36–75). Transportation Research Board Special Report No. 247, Washington, DC: National Academy Press.

Dall'Asta, E., & Roncella, R. (2014). A comparison of semiglobal and local dense matching algorithms for surface reconstruction. *International Archives of the Photogrammetry, Remote Sensing and Spatial Information Sciences, 40*(5), 187–194.

De Agostino, M., Lingua, A., & Piras, M. (2012). Rock face surveys using a LiDAR MMS. *Italian Journal of Remote Sensing, 44*, 141–151.

Debella-Gilo, M., & Kääb, A. (2012). Measurement of surface displacement and deformation of mass movements using Least Squares Matching of repeat high resolution satellite and aerial images. *Remote Sensing, 4*, 43–67.

Delacourt, C., Allemand, P., Casson, B., & Vadon, H. (2004). Velocity field of the 'La Clapiere' landslide measured by the correlation of aerial and QuickBird satellite images. *Geophysical Research Letters, 31*, paper No. 15619.

Delacourt, C., Allemand, P., Berthier, E., Raucoules, D., Casson, B., Grandjean, P., et al. (2007). Remote-sensing techniques for analysing landslide kinematics: A review. *Bulletin de la Société Géologique de France, 178*(2), 89–100.

Dermanis, A. (2011). Fundamentals of surface deformation and application to construction monitoring. *Applied Geomatics, 3*(1), 9–22.

Dewez, T. J. B. (2014). Reconstructing 3D coastal cliffs from airborne oblique photographs without ground control points. *ISPRS Annals of the Photogrammetry, Remote Sensing and Spatial Information Sciences, 2*(5), 113–116.

Eisenbeiss, H., & Sauerbier, M. (2012). Investigation of UAV systems and flight modes for photogrammetric applications. *The Photogrammetric Record, 26*(136), 400–421.

Fastellini, G., Radicioni, F., & Stoppini, A. (2011). The Assisi landslide monitoring: A multi-year activity based on geomatic techniques. *Applied Geomatics, 3*(2), 91–100.

Feng, T., Liu, X., Scaioni, M., Lin, X., & Li, R. (2012). Real-time landslide monitoring using close-range stereo image sequences analysis. In *Systems and Informatics (ICSAI), 2012 International Conference on* (ICSAI 2012) (pp. 249–253), Yantai, P.R. China, May 19–21, 2012.

Förstner, W., Gülch, E. (1987, June). A fast operator for detection and precise location of distinct points, corners and centres of circular features. In *Proceedings of ISPRS Intercommission Conference on Fast Processing of Photogrammetric Data*, (pp. 281–305), Interlaken, Switzerland.

Fraser, C. S. (1996). Network design. In K. B. Atkinson (Ed.), *Close range photogrammetry and machine vision* (pp. 256–281). Dunbeath, Caithness, Scotland, UK: Whittles Publishing.

Fraser, C. S., Woods, A., & Brizzi, D. (1996). Hyper redundancy for accuracy enhancement in automated close range photogrammetry. *The Photogrammetric Record, 20*, 205–217.

Fraser, C. S. (2013). Automatic camera calibration in close range photogrammetry. *Photogrammetric Engineering and Remote Sensing, 79*, 381–388.

Froese, C. R., & Moreno, F. (2014). Structure and components for the emergency response and warning system on Turtle Mountain, Alberta, Canada. *Natural Hazards, 70*(3), 1689–1712.

Furukawa, Y., & Ponce, J. (2010). Accurate, dense, and robust multi-view stereopsis. *IEEE Transactions on Pattern Analysis and Machine Intelligence, 32*(8), 1362–1376.

Ghuffar, S., Székely, B., Roncat, A., & Pfeifer, N. (2013). Landslide displacement monitoring using 3D range flow on airborne and terrestrial LiDAR data. *Remote Sensing, 5*(6), 2720–2745.

Grün, A. (1985). Adaptive least squares correlation: a powerful image matching technique. *South African Journal of Photogrammetry, Remote Sensing and Cartography, 14*, 175–187.

Grün, A., & Baltsavias, E. P. (1988). Geometrically constrained multiphoto matching. *Photogrammetric Engineering and Remote Sensing, 54*(5), 633–641.

Grün, A. (2012). Development and status of image matching in photogrammetry. *The Photogrammetric Record, 27*, 36–57.

Gu, Z., Feng, T., Scaioni, M., Wu, H., Liu, J., Tong, X., & Li, R. (2014). Experimental results of elevation change analysis in the Antarctic Ice sheet using DEMs from ERS and ICESat data. *Annals of Glaciology, 55*(66), 198–204.

Guidi, G., Gonizzi, S., & Micoli, L. L. (2014). Image pre-processing for optimizing automated photogrammetry performances. *ISPRS Annals of The Photogrammetry, Remote Sensing and Spatial Information Sciences, 2*(5), 145–152. doi:10.5194/isprsannals-II-5-145-2014.

Haala N (2013) The landscape of dense image matching algorithms. In *Proceedings of Photogrammetric Week 2013, Stuttgart, Germany* (pp. 271–284).

Hartley, R., & Zissermann, A. (2006). *Multiple view geometry in computer vision*. UK: Cambridge University Press.

Heritage, G. L., & Large, A. R. G. (2009). *Laser scanning for the environmental sciences* (p. 302). Chichester, UK: Wiley.

Hirschmüller, H. (2005). Accurate and efficient stereo processing by semi-global matching and mutual information. In *Proceedings of Conference on Computer Vision and Pattern Recognition (CVPR'05)*, (p. 8). *San Diego, CA, USA*, June 20–26, 2005.

Hoffmann, C. M. (1989). *Geometric and solid modeling: An introduction*. San Francisco: Morgan Kaufmann Publishers Inc.

Hong, Y., He, X., Cerato, A., Zhang, K., Hong, Z., & Liao, Z. (2014). Predictability of a physically-based model for rainfall-induced Shallow Landslides: Model development and case studies. In M. Scaioni (Ed.), *Modern technologies for landslide investigation and prediction* (pp. 165–178). Berlin, Heidelberg: Springer.

Hungr, O., Leroueil, S., & Picarelli, L. (2014). The Varnes classification of landslide types, an update. *Landslides, 11*, 167–194.

Intrieri, E., Gigli, G., Mugnai, F., Fanti, R., & Casagli, N. (2012). Design and implementation of a landslide early warning system. *Engineering Geology, 147–148*, 124–136.

Jaboyedoff, M., Oppikofer, T., Abellán, A., Derron, M. H., Loye, A., Metzger, R., & Pedrazzini, A. (2012). Use of LIDAR in landslide investigations: A review. *Natural Hazards, 61*, 1–24.

Jazayeri, I., & Fraser, C. S. (2010). Interest operators for feature-based matching in close range photogrammetry. *The Photogrammetric Record, 25*(129), 24–41.

Le Moigne, J., Netanyahu, N. S., & Eastman, R. D. (2011). *Image registration for remote sensing* (p. 484). UK: Cambridge University Press.

LePrince, S., Berthier, E., Ayoub, F., Delacourt, C., & Avouac, J. P. (2008). Monitoring earth surface dynamics with optical imagery. *Eos Transactions, 89*, 1–5.

Li, Z., & Grün, A. (2004). Automatic DSM generation from linear array imagery data. *International Archives of The Photogrammetry, Remote Sensing and Spatial Information Sciences, 35*(B3), 128–133.

Lindenbergh, R., Pfeifer, N. (2005). A statistical deformation analysis of two epochs of terrestrial laser data of a lock. In *Proceedings of 7th Conference on 'Optical 3-D Measurement Techniques'*, (Vol. 2, pp. 61–70). *Vienna*, October 3–5, 2005.

Lindenbergh, R. (2010). Chapter 7—Engineering applications. In G. Vosselman & H. G. Maas (Eds.), *Airborne and terrestrial laser scanning* (pp. 237–270). Boca Raton, FL, USA: Taylor and Francis Group.
Lowe, D. G. (2004). Distinctive image features from scale-invariant keypoints. *International Journal of Computer Vision, 60*(2), 91–110.
Luhmann, T. (2009). Precision potential of photogrammetric 6DOF pose estimation with a single camera. *ISPRS Journal of Photogrammetry and Remote Sensing, 64*(3), 275–284.
Luhmann, T., Robson, S., Kyle, S., & Böhm, J. (2013). *Close range photogrammetry: 3D imaging techniques* (p. 702). Germany: Walter De Gruyter Inc.
Maas, H. G. (1996). Automatic DEM generation by multi-image feature based matching. *International Archives of Photogrammetry and Remote Sensing, 31*(3), 484–489.
Mantovani, F., Soeters, R., & van Westen, C. J. (1996). Remote sensing techniques for landslide studies and hazard zonation in Europe. *Geomorphology, 15*(2), 213–225.
Mazzanti, P., Brunetti, A., & Bretschneider, A. (2014). A new approach based on terrestrial remote sensing techniques for rock fall hazard assessment. In M. Scaioni (Ed.), *Modern technologies for landslide investigation and prediction* (pp. 69–87). Berlin, Heidelberg: Springer.
Metternicht, G., Hurni, L., & Gogu, R. (2005). Remote sensing of landslides: An analysis of the potential contribution to geo-spatial systems for hazard assessment in mountainous environments. *Remote Sensing of Environment, 98*, 284–303.
Monserrat, O., Moya, J., Luzi, G., Crosetto, M., Gili, J. A., & Corominas, J. (2013). Non-interferometric GB-SAR measurement: application to the Vallcebre landslide (eastern Pyrenees, Spain). *Natural Hazards Earth System Science, 13*, 1873–1877.
Motta, M., Gabrieli, F., Corsini, A., Manzi, V., Ronchetti, F., & Cola, S (2013). Landslide displacement monitoring from multi-temporal terrestrial digital images: Case of the Valoria Landslide site. In Margottini et al. (Eds.), *Landslide science and practice* (Vol. 2, pp. 73–78). Berlin, Heidelberg: Springer.
Nex, F., & Remondino, F. (2014). UAV for 3D mapping applications. *Applied Geomatics, 6*(1), 1–15.
Niethammer, U., James, M. R., Rothmund, S., Travelletti, J., & Joswig, M. (2012). UAV-based remote sensing of the Super-Sauze landslide: Evaluation and results. *Engineering Geology, 128*, 2–11.
Ohnishi, Y., Nishiyama, S., Yano, T., Matsuyama, H., & Amano, K. (2006). A study of the application of digital photogrammetry to slope monitoring systems. *International Journal of Rock Mechanics and Mining Sciences, 43*, 756–766.
Pears, N., Liu, Y., & Bunting, P. (2012). *3D Imaging, Analysis and Applications* (p. 499). London: Springer.
Pirotti, F., Guarnieri, A., & Vettore, A. (2013). State of the art of ground and aerial laser scanning technologies for high-resolution topography of the earth surface. *European Journal of Remote Sensing, 46*, 66–78.
Pirotti, F., Guarnieri, A., Masiero, A., Gregoretti, C., Degetto, M., & Vettore, A. (2014). Micro-scale landslide displacements detection using Bayesian methods applied to GNSS data. In M. Scaioni (Ed.), *Modern technologies for landslide investigation and prediction* (pp. 123–138). Berlin, Heidelberg: Springer.
Pomerleau, F., Colas, F., Siegwart, R., & Magnenat, S. (2013). Comparing ICP variants on real-world data sets. *Autonomous Robots, 34*(3), 133–148.
Previtali, M., Barazzetti, L., Scaioni, M., & Tian, Y. (2011). An automatic multi-image procedure for accurate 3D object reconstruction. In *Proceedings of 4th International Congress on Image and Signal Processing (CISP), Shanghai* (Vol. 3, pp. 1400–1404), October 15–17, 2011.
Previtali, M., Barazzetti, L., & Scaioni, M. (2014). Accurate 3D surface measurement of mountain slopes through a fully automated imaged-based technique. *Earth Science Informatics, 7*(2), 109–122.
Qiao, G., Lu, P., Scaioni, M., Xu, S., Tong, X., Feng, T., Wu, H., Chen, W., Tian, Y., Wang, W., & Li, R. (2013). Landslide investigation with remote sensing and sensor network: From susceptibility mapping and scaled-down simulation towards in situ sensor network design. *Remote Sensing, 5*(9), 4319–4346. doi:10.3390/rs5094319.

Raguse, K., & Heipke, C. (2009). Synchronization of image sequences—a photogrammetric method. *Photogrammetric Engineering and Remote Sensing, 75*(4), 535–546.
Remondino, F., & Stoppa, D. (2013). *TOF range-imaging cameras*. Berlin, Heidelberg: Springer.
Remondino, F., Spera, M. G., Nocerino, E., Menna, F., & Nez, F. (2014). State of the art in high density image matching. *The Photogrammetric Record, 29*(146), 144–166.
Roncella, R., Scaioni, M., & Forlani, G. (2004). Application of digital photogrammetry in geotechnics. *International Archives of the Photogrammetry, Remote Sensing and Spatial Information Sciences, 35*(B/V), 93–98.
Roncella, R., & Forlani, G. (2014). A fixed terrestrial photogrammetric system for landslide monitoring. In M. Scaioni (Ed.), *Modern technologies for landslide investigation and prediction* (pp. 43–67). Berlin, Heidelberg: Springer.
Roncella, R., Romeo, E., Barazzetti, L., Gianinetto, M., & Scaioni, M. (2012). Comparative analysis of digital image correlation techniques for in-plane displacement measurements. In *Proceedings of 5th International Congress on Image and Signal Processing (CISP)* (pp. 721–726). *Chongqing, P.R. China*, October 16–18, 2012.
Rosenfeld, A., & Kak, A. C. (1976). Digital picture processing (Vol. 1). Elsevier.
Rosu, A. M., Pierrot-Deseilligny, M., Delorme, A., Binet, R., & Klinger, Y. (2014). Measurement of ground displacement from optical satellite image correlation using the free open-source software MicMac. *ISPRS Journal of Photogrammetry and Remote Sensing,* doi:10.1016/j.isprsjprs.2014.03.002.
Scaioni, M. (2013). Remote sensing for landslide investigations: From research into practice. *Remote Sensing, 5*(11), 5488–5492.
Scaioni, M., Roncella, R., & Alba, M. I. (2013a). Change detection and deformation analysis in point clouds: Application to rock face monitoring. *Photogrammetric Engineering and Remote Sensing, 79*(5), 441–456.
Scaioni, M., Lu, P., Chen, W., Qiao, G., Wu, H., & Feng, T., et al. (2013b). Analysis of spatial sensor network observations during landslide simulation experiments. *European Journal of Environmental and Civil Engineering, 17*(9), 802–825.
Scaioni, M., Tong, X., & Li, R. (2013c). Application of GLAS laser altimetry to detect elevation changes in East Antarctica. *ISPRS Annals of The Photogrammetry, Remote Sensing and Spatial Information Sciences, 2*(5/W2), 253–258.
Scaioni, M., Feng, T., Barazzetti, L., Previtali, M., Lu, P., & Giao, G., et al. (2014a). Some applications of 2D and 3D photogrammetry during laboratory experiments for hydrogeological risk assessment. *Geomatics, Natural Hazards and Risk,* doi:10.1080/19475705.2014.885090.
Scaioni, M., Longoni, L., Melillo, V., & Papini, M. (2014b). Remote sensing for landslide investigations: An overview on recent achievements and perspectives. *Remote Sensing, 6*(10), 9600–9652. doi:10.3390/rs6109600.
Scaioni, M., Feng, T., Barazzetti, L., Previtali, M., & Roncella, R. (2014c). Image-based deformation measurement. *Applied Geomatics*, doi:10.1007/s12518-014-0152-x.
Scharstein, D., & Szeliski, R. (2002). A taxonomy and evaluation of dense two-frame stereo correspondence algorithms. *International Journal of Computer Vision, 47*(1–3), 7–42.
Seitz, S. M., Curless, B., Diebel, J., Scharstein, D., & Szeliski, R. (2006). A comparison and evaluation of multi-view stereo reconstruction algorithms. In *Proceedings of Conference on Computer Vision and Pattern Recognition (CVPR), New York* (Vol. 1, pp. 519–526), June 17–22, 2006.
Spencer, L., & Shah, M. (2004). Temporal synchronization from camera motion. In *Proceedings of 6th Asian Conference on Computer Vision* (Vol. 1, pp. 515–520). *Jeju Island, Korea*, January 27–30, 2004.
Stumpf, A., Malet, J. P., Allemand, P., & Ulrich, P. (2014). Surface reconstruction and landslide displacement measurements with Pléiades satellite images. *ISPRS Journal of Photogrammetry and Remote Sensing, 95*, 1–12.
Tao, V., & Li, J. (2007). *Advances in mobile mapping technology*. ISPRS Book Series No. 4. London: Taylor & Francis Group.

Teunissen, P. J. G. (2000). *Testing theory: An introduction*. Series on Mathematical geodesy and positioning. The Netherlands: Delft University Press.

Tommaselli, A. M. G., Moraes, M. V. A., Silva, L. S. L., Rubio, M. F., Carvalho, G. J., & Tommaselli, J. T. G. (2014). Monitoring marginal erosion in hydroelectric reservoirs with terrestrial mobile laser scanner. *International Archives of the Photogrammetry, Remote Sensing and Spatial Information Sciences, 40*(5), 589–596.

Toschi, I., Capra, A., De Luca, L., Beraldin, J. A., & Cournoyer, L. (2014). On the evaluation of photogrammetric methods for dense 3D surface reconstruction in a metrological context. *ISPRS Annals of the Photogrammetry, Remote Sensing and Spatial Information Sciences, 2*(5), 371–378.

Travelletti, J., Delacourt, C., Allemand, P., Malet, J. P., Schmittbuhl, J., Toussaint, R., & Bastard, M. (2012). Correlation of multi-temporal ground-based optical images for landslide monitoring: Application, potential and limitations. *ISPRS Journal of Photogrammetry and Remote Sensing, 70*, 39–55.

Wallis, R. (1976). An approach to the space variant restoration and enhancement of images. In *Proceeding of Symposium on Current Mathematical Problems in Image Science, Naval Postgraduate School* (pp. 329–340). *Monterey, CA, USA*. November 11–12, 1976.

Wasowski, J., & Bovenga, F. (2014). Investigating landslides and unstable slopes with satellite Multi Temporal Interferometry: Current issues and future perspectives. *Engineering Geology, 174*, 103–138.

Wiegand, C., Rutzinger, M., Heinrich, K., & Geitner, C. (2013). Automated extraction of shallow erosion areas based on multi-temporal ortho-imagery. *Remote Sensing, 5*, 2292–2307.

Wu, J., Gilliéron, P. Y., & Merminod, B. (2012). Cell-based automatic deformation computation by analyzing terrestrial LIDAR point clouds. *Photogrammetric Engineering and Remote Sensing, 78*, 317–329.

Wujanz, D., Krüger, D., & Neitzel, F. (2013a). Defo scan++: Surface based registration of terrestrial laser scans for deformation monitoring. In *Proceedings of 2nd Joint International Symposium on Deformation Measurement* (*JISDM*), *Nottingham* (p. 7), September 2–6, 2013.

Wujanz, D., Neitzel, F., Hebel, H. P., Linke, J., & Busch, W. (2013b). Terrestrial radar and laser scanning for deformation monitoring: First steps towards assisted radar scanning. *ISPRS Annals of Photogrammetry, Remote Sensing and Spatial Information Sciences*, 2(5/W2), 325–330.

Xue, Q., Zhang, M., Zhu, L., Cheng, X., Pei, Y., & Bi, J. (2014). Quantitative deformation analysis of landslides based on multi-period DEM data. In K. Sassa et al. (Eds.), *Landslide science for a safer geoenvironment* (Vol. 2, pp. 201–207). Berlin, Heidelberg: Springer.

Zhang, L. (2005). Automatic digital surface model (DSM) generation from linear array images. *Ph. D dissertation, Mitteilungen Nr. 90* (p. 199). Institute of Geodesy and Photogrammetry, ETH Zurich.

Zhao, H., Zhang, B., Wu, C., Zuo, Z., Chen, Z., & Bi, J. (2014). Direct georeferencing of oblique and vertical imagery in different coordinate systems. *ISPRS Journal of Photogrammetry and Remote Sensing, 95*, 122–133.

A Fixed Terrestrial Photogrammetric System for Landslide Monitoring

Riccardo Roncella and Gianfranco Forlani

Abstract Though landslide alert is based on monitoring systems capable of high frequency, highly accurate, continuous-operation photogrammetry has been used since long time to periodically control the evolution of landslides. In this chapter, a fixed terrestrial stereo photogrammetric system is presented. It has been developed to monitor landslides and, in general, changes in digital surface model (DSM) of the scene framed by the cameras. The system is made of two single-lens reflex (SLR) cameras, each contained in a sealed box and controlled by a computer that periodically shoots an image and sends it to a host computer. Once an image pair is received, the DSM of the scene is generated by digital image correlation on the host computer and made available for archiving or analysis. The system has been installed and is being tested on the Mont de la Saxe landslide, where several monitoring systems are active and provide reference data. Instability of the camera attitude has been noticed and corrected with an automated procedure by image resampling. First comparisons with interferometric synthetic aperture radar data show a good agreement of the displacements over time.

Keywords Hazard · Landslides · Terrestrial · Photogrammetry · Monitoring · Automation · Algorithms

R. Roncella (✉) · G. Forlani
Department of Civil, Environmental, Land Management Engineering and Architecture, University of Parma, via Parco Area delle Scienze, 181/a, 43124 Parma, Italy
e-mail: riccardo.roncella@unipr.it

G. Forlani
e-mail: gianfranco.forlani@unipr.it

M. Scaioni (ed.), *Modern Technologies for Landslide Monitoring and Prediction*,
Springer Natural Hazards, DOI 10.1007/978-3-662-45931-7_3

1 Introduction

Landslide surveillance and monitoring systems face a large variety of cases and situations. Indeed, the behaviour of a landslide (in the time domain as well as in the space domain) depends on multiple factors, which each time challenge geologists and surveyors in a different way.

In principle, the main design parameter when installing a measurement and monitoring system is the accuracy needed to assess with a given probability the magnitude of the expected displacements and deformations.

However, a number of other issues influence the choice of the best monitoring system to implement the size of the area to control, the frequency of data acquisition, the time to deliver the results (alert time), the stability of the reference system, the influence of atmospheric parameters on measurement accuracy or operation, the site constraints and limitations for the installation of the system (visibility, energy consumption, data telemetry, safety, security), etc.

No single measurement system is therefore suited for any situations, and the selection of the best solution for a specific site is not always straightforward; often, if the risk is high and the budget available allows for, more systems are integrated into sensor networks (see Qiao et al. 2013).

Specific *geotechnical instrumentations* (such as inclinometers and piezometers) aim at controlling directly or indirectly the movements of the slip surface (Angeli et al. 2000); other instruments such as wire *extensometers*, as reported in Corominas et al. (2000), can be installed to sense and measure displacements between specific points. Displacement and movements of terrain are mainly controlled by *topographic* and *photogrammetric surveys*. Traditionally, the survey rests on the accurate determination of the position of a relatively few points (a few dozen at most) distributed over the area under control; the displacement between two measurement epochs is obtained by coordinate differences. Recent sensors and processing techniques allow rather for the determination of the entire terrain surface at any measurement epoch. In this case, more than a measure of surface deformation can be used. For instance, only the vertical component of the displacement can be considered (e.g. by *rasterization* of the elevations at a suitable resolution); rather, surface-to-surface distance as discussed in Cignoni et al. (1998) and in Aspert et al. (2002) can be evaluated.

Robotic total stations have reached a very high level of automation and can control with dozens of prisms with high frequency (see Bertacchini et al. (2011)); the accuracy level, according to specifications, is below in the mm level for short distances, while refraction is the main limiting factor for distances of several hundred metres. Monitoring capabilities (continuous operation and data transmission) are available.

More recently, *terrestrial laser scanning* (TLS) and *airborne laser scanning* (ALS) are also being used as alternative, the former particularly with landslides on steep slopes (see Bitelli et al. (2004), Jaboyedoff et al. (2012), Abellán et al. (2014)). However, normally TLS is used only for periodic surveys (i.e. they are not continuously operating); in large landslides, where more than one scan station might be necessary, coregistration is needed as in Barbarella and Fiani (2013).

Ground-based or *spaceborne InSAR* is also being used to monitor landslides, provided that the type of terrain allows for good coherence between multitemporal images and the main displacement component is along the line of sight (see Monserrat et al. (2014), Wasowski and Bovenga (2014), Scaioni et al. (2014a)). The ground-based version needs a stable installation, which results in skyrocketing the monitoring costs. However, it provides almost continuous monitoring over a large area with millimetre sensitivity (Crosetto et al. 2014).

Photogrammetry has been used since long to control the evolution of landslides, either from aerial images as in Casson et al. (2003) or from ground, as reported in Mora et al. (2003) and in Cardenal et al. (2008). In the analytical photogrammetry era, detection of the movement was based on evaluation of digital surface model (DSM) changes (measured manually along profiles) or on displacement of targets distributed over the area. Today, automatic block orientation and DSM generation, with resolutions and accuracies comparable to TLS surveys, make photogrammetry an alternative worth considering (see Scaioni et al. (2014b)).

In the above-mentioned cases, photogrammetry is not used for proper monitoring and alert, since the surveys are executed on a regular basis (e.g. monthly or annual) in order to assess whether a slow movement is still active. Ground control points (GCP's), located outside the landslide area, must be used to maintain a stable reference system. In this respect, photogrammetry differs from the other monitoring techniques, where the sensor is located on a stable monument (together with orientation points if needed) to ensure invariance of the reference system: images at different epochs may be taken from different locations.

Under certain conditions, however, a fixed photogrammetric system, capable to acquire the DSM of the landslide body with high frequency, may provide a continuous and low-cost flow of dense geometric data, even if inherent limits on accuracy and camera stability should be dealt with. Even if fixed and continuously operating, a photogrammetric system cannot perform as *landslide early warning system* (see Intrieri et al. 2013), because of its limitation to daylight and good weather conditions; however, its strength lies in low cost, simplicity of components and scalability.

1.1 Previous Work on the Topic

The photogrammetric system developed and applied to the Mont de la Saxe landslide (see Sect. 6) is to our best knowledge the first attempt to use stereo photogrammetry from outdoor permanent photo stations for monitoring purposes. Two monocular systems with similar characteristics are described in Travelletti et al. (2012) and in Motta et al. (2013). In the former, a Nikon D70 camera with a 50-mm lens is placed on a pillar facing the landslide, at a distance ranging from 300 to 900 m; four images per day are acquired around noon. The exterior orientation

(EO) of the camera is determined by space resection (see Luhmann et al. 2013) from several GCP's measured with Global Navigation Satellite System (GNSS). The processing software tracks features in consecutive images producing a projection of the displacement field of the landslide in the image plane camera. To estimate the actual displacement in ground coordinates, a high-resolution light detection and ranging (LiDAR) DSM of the area is projected on the image and associated with the image coordinates. Comparison of the system with GNSS monitoring over a period of about 2 years shows good correlation, and a relative accuracy between 10 and 20 % is reported. Systematic errors due to changes in the DSM morphology as well as from camera movement are also reported; the latter can be detected by a statistical test on the estimated displacement field. System performance is also strongly affected by illumination conditions, weather conditions and seasonal changes in land surface.

The stability of the reference system is a key issue with any monitoring system; optical geodetic measurements with collimators, theodolites and total stations traditionally rely on precise centering devices and stable reference points for orientation; this applies to laser scanning as well (see, for instance, Monserrat and Crosetto (2008) and Scaioni et al. (2013)). For a fixed photogrammetric system, in principle, there are two options available: if the imaged area contains also stable points, the reference system can be defined by means of such points and cameras can be oriented each time; this can be considered, however, rather an exception than a rule. Normally, therefore, the camera must be placed on a stable monument and camera orientation must be performed at system installation and periodically checked.

2 The Photogrammetric System

2.1 Stereo-Camera System Architecture

The *stereo-camera* system is designed to produce three-dimensional models of the scene with a scheduled frequency of acquisition and processing. Being conceived for installation in high mountain environment, in hard-to-reach locations, the hardware is remotely controlled and checked, to keep to a minimum the on-site user intervention.

To keep system costs down, customer-grade cameras and off-the-shelf hardware have been used. A single-lens reflex (SLR) camera with interchangeable lens coupled with a programmable board for camera control and data transmission is used; in the current configuration, a 21-Mpix Canon EOS 5D Mark II with a 50-mm lens is employed in each unit.

A microcomputer controls shooting, storing, processing and data transmission. The operating system is a lightweight Linux Debian 6.0 'Squeeze', with a plenty of available open-source software programs. The microcomputer is connected to a Universal Mobile Telecommunications System (UMTS) modem for data transmission. The user through secure shell (SSH) protocol controls all the acquisition parameters, as well as battery status, temperatures, etc. The system is permanently connected to the Internet (it reconnects automatically in case of loss of GSM signal due, e.g., to temporary hardware malfunction, low batteries, etc.). Every 15 min, the system clock is updated to have a good synchronization between the two acquisition units. Anyway, the slow velocity of the landslide displacements does not result in any synchronization problem (see Raguse and Heipke (2009)).

Image processing is completely automatic: at each measurement epoch, both cameras send data to a centralized archiving and processing server, which autonomously produce a DSM of the slope. Ideally, the new DSM should be automatically compared with a reference DSM of the slope to evaluate the movements and, in case, send an alert: this feature is not yet implemented, however.

Everything is housed in an IP67 (International Protection class) watertight box (see Fig. 1) easy to carry, sufficiently robust to endure critical weather conditions, equipped with a special joint for easy pointing of the camera. To prevent condensation at low temperatures, the air is removed with a vacuum pump from a valve.

Operating outdoor temperatures well below −20 °C and exceeding +30 °C can be reached; as far as energy supply is concerned, a 50-W solar panel and a pack of three 17-Ah batteries ensure continuous operation. Batteries are continuously recharged by the solar panel. In the case of persistent cloud coverage, the autonomy of the system is about 3 days. If available at a convenient distance, connection to a power outlet is possible.

Fig. 1 *Left* the watertight IP67 box housing the system. *Right* one of the system units and the solar panel installed on a pole

2.2 System Calibration

In order to operate, the stereo-camera system needs to be calibrated. Two sets of calibration parameters must be estimated: the *interior orientation* (IO) (including lens distortion) parameters and the *exterior orientation* (EO) parameters of each camera (see Luhmann et al. (2013)).

2.2.1 Interior Orientation Parameters

Depending on the situation on the installation site and on the instrumentation available, several calibration techniques can be used. In principle, if enough control points can be measured on the landslide surface, a self-calibrating bundle block adjustment including also automatically extracted tie points could be carried out; with this method, all relevant parameters can be estimated in a single step. However, due to safety reasons and/or site accessibility, in most cases, proper targets cannot be installed to be used as GCP's, which must be selected on natural features. Moreover, being in this case the photogrammetric block made of only two images, strong correlations between and within the calibration parameters are expected, especially if the scene depth is limited. If such correlations between the two sets are to be reduced, a more reliable *two-stage* calibration is instead accomplished, one for IO and the second for EO parameters.

The former stage is in principle a standard photogrammetric calibration task, to be performed with a strong camera network and well-defined targets; to this aim, a laboratory calibration test field can be used. However, to account for the protective front glass of the box, the calibration images must be shot with the camera inside the box, remotely controlled via Ethernet cable. Due to the long focal length optics, a relatively large calibration panel is necessary: rotating and moving the panel in different positions and also changing the box pose provide camera stations with wide base lengths; rotating the box around the camera optical axis should reduce correlations between IO and EO parameters.

In the first calibration executed on the full-frame Canon EOS 5D Mark II camera with a 50-mm lens, a bundle block full-field analytical calibration was performed with 20 images approximately distributed on a spherical surface centred on the calibration panel. A root mean square (RMS) of the collinearity residuals of about 0.7 pixels for both cameras was obtained. Usually, with FX (full-frame) consumer-grade camera and optics of the same quality, better results should be expected (in the order of 0.2–0.3 pixels), and as a matter of fact, the calibration of the same camera without the box provided much better residuals. The most obvious reason is that the thick protective glass introduces some distortion not modelled by the calibration procedure [which is based on a standard distortion model (see Brown 1971)]. Currently, a thinner glass has been installed, trying to reduce distortion effects and obtain brighter images. In the correlation matrix, no value exceeds 50 %, apart from the principal point position with respect to the decentring distortion parameters that always show high correlations.

Good accuracy and low residual correlations between IO parameters must be obtained, since the EO procedure cannot make up for deficiencies, as it happens to some extent with block orientation performed with GCP's. As in direct georeferencing from mobile mapping, any error in IO parameters would be reflected in this case directly on the ground coordinates of the points.

2.2.2 Exterior Orientation Parameters

There are several ways to determine the EO of the two cameras. In principle, determining the relative orientation of the image pair and providing a good approximation of the base length are enough to set the system: indeed, what matters is computing a displacement field and this can be done in an arbitrary reference system. In practice, especially if more sensors are used, it is desirable to have the cameras georeferenced in a mapping system.

As already noticed above, using artificial or natural GCP's located on the landslide body is often difficult. The area covered by the two cameras will be in many cases completely in the active part of the landslide. This means that GCP's can be used just once and that they must be measured when the EO parameters of the cameras are determined; using targets would therefore be unjustified. Optimal natural targets, however, are difficult to find because they should be well defined in both the images and the theodolite eyepiece. Though today's theodolites are equipped with long-range no-prism rangefinders, their accuracy might not be enough and depends also on rock surface colour and characteristics, on the size of the footprint of the laser and on its incidence angle with the slope. On the other hand, determination of the points by angular intersection from two theodolite stations requires intervisibility and is error prone due to perspective differences between the two sightings.

Of course, whenever the site characteristics allow for it, using a theodolite to determine GCP's for the orientation is still the best option to take. However, a combination of topographic and photogrammetric information can also be used to determine the EO parameters of both cameras that can be considered a case of integrated sensor orientation procedure (Jacobsen 2004) as used in the early days of GNSS-assisted aerial triangulation.

On top of the box, a 5/8″ screw can host a retro-reflective prism (or a GNSS antenna). The prism is eccentric with respect to the camera-perspective centre by an unknown eccentricity vector e.

With the two boxes installed in front of the landslide, taking into account the eccentricity and measuring with a total station the prism position, the coordinates of the projection centre of each camera station can be obtained; the attitude parameters are obtained indirectly using tie points in a bundle adjustment. If three not-aligned cameras were used, the information (prism locations' eccentricity vector and tie point image coordinates) would be sufficient to provide the EO of the cameras. The mathematical model is the same used in aerial triangulation (Forlani and Pinto 1994): each prism location is introduced as pseudo-observation in the bundle adjustment, together with the collinearity equations:

$$X_p = X_0 + \mathbf{R}_c^g e \tag{1}$$

where X_P is the prism position, X_0 is the (unknown) perspective centre and $\mathbf{R}_c^g$ is the (unknown) rotation matrix from image to object space.

Since the photogrammetric monitoring system consists of just two acquisition boxes, the normal matrix of the bundle adjustment would be in this case rank deficient unless (at least) a GCP is provided. Indeed, the general rotation around the base length cannot be determined, as would be in the symmetric relative orientation. Therefore, a few GCP's must be measured on the landslide; however, they need not to be well distributed, as it would be necessary if the space resection were used.

The eccentricity vector e can be determined in image space by a simple calibration procedure (Forlani and Pinto 2007). To this aim, one or more images of a test field with GCP's are taken from the box with the prism mounted on the top. At each shooting, the prism position is measured with a theodolite. After image acquisition, the orientation of the images is accomplished, e.g., by space resection over the GCP's, providing the perspective centres. The components of the eccentricity vector in object space at each shooting can therefore be computed; multiplication by the attitude matrix brings it in image space. Taking a few images allows us to estimate the accuracy of the vector components, normally in the order of 1–2 mm with a total station. In the case a GNSS antenna is used, the accuracy is generally lower (1–2 cm), and the GNSS position should be referred to the phase centre and not to the mechanical reference plane, unless the antenna is levelled.

3 Metrological Analysis of the System

The system described so far, in order to be correctly employed in a landslide or rockfall monitoring scenario, requires a deep analysis of theoretical accuracy that can be achieved according to the imaging geometry and the hardware features. As a very simple rule of thumb, the accuracy of a generic stereo- or multi-image imaging system is proportional to the mean image scale or the mean ground sampling distance (GSD) of the camera, as a very straightforward error propagation procedure can show. For instance, as far as a simple stereo-camera system in a normal case geometry is considered, the error along the depth/optical axis direction (the error ellipse presents usually its major axis along such direction) is

$$\sigma_i = m_B \frac{Z}{B} \sigma_{im} \tag{2}$$

where m_B is the average image scale, σ_{im} is the measurement precision of image coordinates and Z/B represents the geometric ratio between the distance from the object (Z) and the base length (B) between the two camera centres; the ratio is usually in the range between 1 and 10.

On the other end, if a more general, multi-image, convergent system is employed, according to Fraser (1996), the precision (σ_i) is

$$\sigma_i = m_B \frac{q}{\sqrt{n}} \sigma_{im} \tag{3}$$

where n is the average number of exposures at each station and q is a design factor expressing the strength of the camera network (basically dependent on the angles between intersecting homologous rays) that may range between 0.4 and 2. If the intersection between the rays is nearly 90°, the precision is the same in all directions. In other words, the use of convergent images results in a better spatial intersection of corresponding rays which may provide a smaller and isotropic error ellipse for the reconstructed point. Convergent poses are then preferable since higher intersecting angles can be obtained; on the other hand, if automatic image matching techniques are used, higher perspective changes can make the matching process less accurate: in other words, σ_{im} in (3) tends to grow as relative angles between the pictures become higher, reducing at some point the precision of the estimated object coordinates. As it can be seen, the precision may be improved by adding more images (n) even if, in that case, the total cost of the monitoring system and the computational demands, accounting more than two images, increase proportionally.

Considering a distance from the object Z of 500 m and a base length B of approx. 100 m, using a slightly convergent stereo-camera system with a 21 Mpix resolution (i.e. a 6.5-μm/pixel FX sensor) equipped with a 50-mm focal lens and considering that, in outdoor good lighting condition, current state-of-the-art area-based matching (ABM) algorithms (see Scaioni et al. 2014b) can achieve ±0.2 pixel accuracy quite systematically, Eq. 2 or Eq. 3 gives a final depth precision of approx. ±6.5 cm.

Equations (2) and (3) give a rough evaluation of expected precision, but assume several parameters to be known; it is worth noting that (2) and (3) consider the error propagation of a generic measurement error of corresponding image points (σ_{im}) on final object coordinates, according to the imaging geometry: such error is usually (and ideally) considered as a zero-mean, Gaussian-distributed, random variable. The resulting object coordinate error can be considered accordingly as a white Gaussian noise, affecting the final restitution. If the imaging system is not stable (e.g. the protecting box has some unwanted movements that change the EO of the system), more systematic and spatially correlated effects on the final restitution should be expected.

External checks relying on independent information should always provide, in the initial and periodical validation of the monitoring system, a direct or indirect evaluation of the results. Anyway, such validation cannot be performed at every single measurement and restitution epoch (or the system would easily become not cost efficient), and the image stability of the system must be enforced and checked automatically.

3.1 Image Stability of the System

A very small not modelled variation in the mutual or individual position or attitude of the camera stations can introduce apparent shifts in the DSM comparison that may well exceed the system accuracy.

The analytical framework that explains the distance error committed by the sensor can be adapted from the general case of model height error in relative orientation as a function of the orientation elements: in that case (see, for instance, Kraus 2008), assuming a pseudo-nadiral geometric configuration, the depth (distance) error dZ is

$$\begin{aligned} dZ = & -\frac{Z}{B}db_x - \frac{(X-B)}{B}db_z - \left(\frac{Z^2}{B} + \frac{X^2}{B}\right)d\phi_1 + \frac{XY}{B}d\varpi_1 - \frac{YZ}{B}d\kappa_1 \\ & + \left(\frac{Z^2}{B} + \frac{(X-B)^2}{B}\right)d\phi_2 - \frac{(X-B)Y}{B}d\varpi_2 + \frac{YZ}{B}dk_2 \end{aligned} \tag{4}$$

where B is the base length between both cameras, X, Y and Z (depth) the object point coordinates and dbx, dbz, $d_{\omega 1}$, $d_{\omega 2}$, $d_{\phi 1}$, $d_{\phi 2}$, $d_{\kappa 1}$, $d_{\kappa 2}$ the relative orientation parameter errors (unknown).

A change in the base length B (i.e. a movement along the direction connecting the two perspective centres) is indeed less likely (usually the support structure of the protection box is extremely rigid and well secured to the ground) and produces little effects on the final restitution: for example, considering the same imaging parameters of the previous section (Z = 500 m, B = 100 m), with a movement db_x of 5 mm, the systematic error introduced in the restitution is approx. 2.5 cm; its amount would therefore not introduce significant changes in the coordinates, if compared to the precision evaluated in the previous section.

On the other hand, it can be easily seen that a rotation error induces an erroneous estimation of the depth that varies quadratically with the object coordinates. For instance, an unaccounted ϕ rotation of 0.01° of just one camera can introduce a variation of more than 40 cm along the optical axis direction.

Conversely, the supporting system is not as impervious to rotational movements: with the snow load on the box in winter and especially the wind thrust, variations to camera attitude angles may occur.

3.2 Analysis of the Sensor Orientation Stability

One of the most important feature the monitoring system should have it's also one of the most difficult to be clearly ensured: as shown in the previous section, an unmodelled variation in the camera pose, or in IO parameters, can easily introduce very high systematic effects that can make the interpretation of the results very hard.

Unfortunately, the influence of climatic conditions on the IO parameters of consumer-grade cameras in close-range photogrammetry is scarcely investigated: all the experiences related by analogy to the one under examination were operated in a temperature-controlled environment that made the camera less prone to thermal changes. Anyway, we must consider that the main purpose of the system is to compare digital models acquired at regular intervals and quite close in time, so the problem could be much more severe than it seems. Most likely, acquiring the images at the same time of day should limit the temperature difference. Equipping the box with temperature sensor and limiting image capture at nearly the same temperature conditions or, alternatively, sending the temperature data to the control station to decide later whether to use or not the images acquired at too different climatic conditions might be an option. Moreover, an abrupt change in focal length, being an effect that is reflected systematically on all the estimated object points, can be easily highlighted (and partially corrected) by looking at stable areas where zero displacements should be expected.

As far as EO parameter stability is concerned, the problem might be even harder to be clearly quantified during the system design phase. A very rigid supporting stage can ensure lower vibration and/or movements of the camera, but, especially in high mountain regions where stronger winds and harder weather conditions should be expected, it can never be taken for granted. At the same time, the structural properties of the supporting stage are usually connected to other practical considerations/decisions that could not always consider as a primary factor the total rigidity of the system: for example, in many installations, the support should also house the solar panel that can exert further forces on the structure; in other circumstances, the support must be light-weighted in order to be easily transportable; sometimes the stage cannot be anchored to a rigid structure (e.g. some concrete structure or a rock mass) and is fixed to the ground by means of some tie rod.

Anyway, trying to quantify at least the order of magnitude of the movements that can affect a system similar to the one described in Sect. 2, the following experiments were conducted: the system installed and tested in the case study illustrated in the next Sect. 6 acquired on a daily basis at least five pictures, one every 2 h, for more than 2 months.

To evaluate the actual rotation of the system at each epoch, it is impossible to provide a direct instrumental measurement of the protecting box with sufficient accuracy to highlight slight pose changes (less than 0.01°). For this reason, the rotation of the camera was estimated at each epoch measuring and comparing corresponding image points. What is currently difficult to quantify, given the nature of the application and the lack of similar experiences in the scientific literature (at least according to author's knowledge), is the interval of time after which the aspect of the slope changed so much that it prevents the identification of enough corresponding points. It was decided to investigate the problem using three different ways to determine the corresponding points:

1. The first method finds automatically corresponding image points with feature-based matching (FBM) algorithms (see Scaioni et al. 2014b);
2. The second consists of manual collimation of recognizable features, well distributed on the frame: it is believed that manual collimation, though less precise, is definitely more robust and ensures reliable recognition of the elements even after a long time interval with respect to the reference frame epoch; and
3. The third method uses ABM algorithms [i.e. adaptive least squares matching—LSM (see Grün 1985)] to improve, where possible, the accuracy of the image coordinates measured manually by the operator.

An initial comparison was made between manual and automatic collimations in terms of number of identified inliers (points that are consistent with respect to a predetermined threshold of acceptance). The first, trivial, outcome is that regardless of the method used for collimation, with the progress of the days, less and less inlier can be extracted reliably for the modification of illumination and shadows on the slope, as well as changes in vegetation and other elements of the slope. Figure 2 shows the number of inliers at different epochs using the FBM method.

It is worth noting that, apart when weather conditions (fog, clouds, haze) prevent any sensible outcome of the algorithm, the higher the inlier number clearly decreases, the higher the difference when images are gathered at different times of the day (see subpicture in Fig. 2). The images should always be compared considering the same time of the day (or better the same sun position) since the algorithms are far from insensitive to illumination and shadow changes.

Therefore, in further analysis, only image pairs where the point matching was reliable were used. Manual collimation is the ideal to provide reference, because the operator is unlikely to make blunders in identifying points.

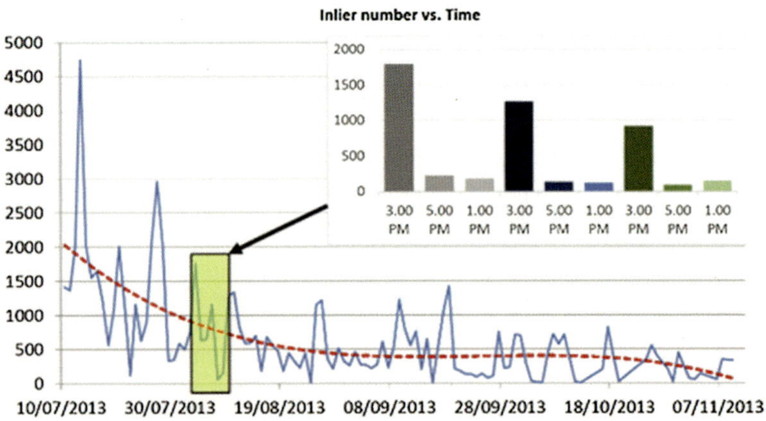

Fig. 2 Inlier number at different epochs using FBM (method 1). On the *upper right* side, the inlier number obtained at different time of the day is reported

At the end of the experiment, the mean value of the system rotation from one epoch to the subsequent was computed: the analysis is obviously just devoted to figure out the order of magnitude of the rotation of the system, while it is rather impossible to guess, just comparing corresponding point positions, the translational part of the movement. In addition, from time to time, a larger rotation of the box has been noticed: whether this is due to a plasticization of the support (unlikely) or a push by external forces (cows, goats and deer usually pasture in the installation area) is not clear; anyway, since such a high movement was found just two times in the series, the data connected to it were considered as blunders and removed from the statistical analysis. The final mean value of the rotation was approx. 0.2° for ω and ϕ components and an order of magnitude lower (approx. 0.02°) for κ angle. At the same time, the repeatability of the rotation estimation was computed: the standard deviations of the angle differences were different, as the imaging geometry could clearly influence the precision in the estimation, for $\omega = 0.004°$, $\phi = 0.002°$ and $\kappa = 0.007°$. Such values can be used to evaluate, correcting the unwanted rotation in some way (see, for instance, the next paragraph), what might be the effect of an incorrect assessment of rotational parameters on the DSM.

For this purpose, a numerical simulation was performed: using the standard deviations of the angle differences, 1,000 error triplets were randomly generated using a Gaussian probability distribution and added to a known initial camera orientation with the same imaging geometry described in the previous paragraph. Five points were considered detected on the surface of the slope at increasing distance, starting from an average distance of 500 m up to a distance of approx. 700 m. For each set of errors, the effect on the coordinates (east, north and height) of each virtual point was evaluated using a forward intersection algorithm. The results are summarized in Table 1. It is worth noticing that for all distances, the percentage of errors in excess of 10 cm on the final model is extremely limited.

3.3 Correction of Unwanted Rotations

To highlight the possible movements between different epochs, and (hopefully) subsequently correcting them, two automatic orientation algorithms have been implemented in the processing pipeline. The first compares each of the two synchronous images obtained at a certain time with the conjugated position of the previous epoch to highlight possible movements. In fact, especially in windy months, this situation occurs almost at every acquisition epoch.

Anyway, since small unwanted variations in the camera attitude are very frequent, the second automatic orientation system corrects, at each epoch, the orientation parameters. The system compares any acquired image with a reference one with known EO and determines the geometric transformation to restore the image plane to the conditions of the original attitude. Assuming that the main contribution to the movement of image plane features is due to unwanted box rotations, a homography transformation can approximately map the reference and the acquired

Table 1 Statistics of errors on DSM coordinates at different distances due to normally distributed camera attitude errors

East (m)				
Distance	RMS	Max	Min	%|err| < 10 cm
500	0.04	0.14	−0.14	98
550	0.05	0.14	−0.15	96
600	0.05	0.15	−0.16	93
650	0.07	0.19	−0.20	87
700	0.08	0.24	−0.25	79
North (m)				
500	0.02	0.07	−0.07	100
550	0.04	0.11	−0.1	100
600	0.05	0.14	−0.13	98
650	0.05	0.14	−0.13	97
700	0.06	0.16	−0.16	93
Height (m)				
500	0.01	0.03	−0.03	100
550	0.01	0.04	−0.05	100
600	0.02	0.06	−0.08	100
650	0.03	0.07	−0.09	100
700	0.03	0.09	−0.12	100

image plane. As shown in Hartley and Zisserman (2006), any two images of a general scene with the same camera centre can be expressed by their *camera matrix* (selecting appropriately the reference system) as follows:

$$\mathbf{P}_1 = \mathbf{K}_1 | \mathbf{I} \quad 0 | \quad \text{and} \quad \mathbf{P}_2 = \mathbf{K}_2 | \mathbf{R} \quad 0 | \tag{5}$$

where $\mathbf{P}_1$ and $\mathbf{P}_2$ represent the *camera matrices*, $\mathbf{K}_1$ and $\mathbf{K}_2$ are the *calibration matrices* (containing the IO parameters associated with each camera) and $\mathbf{R}$ is the *relative rotation matrix* between both cameras. Considering a generic object point in vector X (in homogeneous coordinates):

$$\begin{aligned} x_1 &= \mathbf{P}_1 X = \mathbf{K}_1 | \mathbf{I} \quad 0 | X \\ x_2 &= \mathbf{P}_2 X = \mathbf{K}_2 | \mathbf{R} \quad 0 | X = \mathbf{K}_2 \mathbf{R} | \mathbf{I} \quad 0 | X \end{aligned} \tag{6}$$

denoting with x_1 and x_2 corresponding image point vectors on the first and second frame (in homogeneous coordinates), Eq. 6 represents the generic collinearity equations for the two images. With a trivial analytical step, it is easy to demonstrate the following relation, mapping corresponding points on the two image planes:

$$x_2 = \mathbf{K}_2 \mathbf{R} \mathbf{K}_1^{-1} x_1 = \mathbf{H} x_1 \tag{7}$$

where $\mathbf{H}$ is a 3×3 *homography matrix* that, in projective geometry, represents a general planar projective transformation.

Notice that Eq. 7 can be used here in two different ways: if some variations in IO parameters of the camera occur (in particular, a change in the focal length was considered critical), the full 8 degree-of-freedom (DoF) **H** matrix should be estimated. On the contrary, if the IO parameters are enough stable (i.e. $\mathbf{K}_1 = \mathbf{K}_2$ for the epochs considered), only the 3 DoF rotation matrix **R** could be evaluated.

Even if, at first glance, the former solution (considering all the DoF in Eq. 7 and estimating the full homography matrix, taking into account the possible variation in the IO parameters) sounds very promising and efficient, the results highlight that in such a case, the overparameterized solution system suffers from numerical instability. On the contrary, quite surprisingly, considering all the IO parameters constant in time seems to produce almost satisfactory results. At present, the system trial, with about 13 months of processed and compared data, did not show any issue connected with changes in focal length. Moreover, during the winter season, when wider temperature ranges should be expected, the landslide body is completely covered by snow, and the system is unusable.

With a procedure identical to that used, for example, to produce panoramic or spherical images (see e.g. Barazzetti et al. 2013), an automatic algorithm that compares a series of interest points identified in areas considered stable determines the transformation with robust estimation models such as RANSAC (RANdom SAmple Consensus, see Fischler and Bolles 1981). If just the rotation matrix **R** is estimated in the procedure, assuming the camera calibration matrix $\mathbf{K} = \mathbf{K}_1 = \mathbf{K}_2$ fixed, from Eq. 7, the corresponding homography matrix **H** can be derived. The-image at the epoch considered is then resampled and brought to a condition 'projectively' equivalent to the configuration in which there is no motion lag.

4 Image Processing Workflow

A service running on a host remote server awaits for incoming images from both stations. As far as a synchronous image pair is available, the DSM generation sequence is activated, as described below:

1. The original images are resampled to remove the lens distortion;
2. Well-defined points are extracted on both images with the speeded-up robust feature (SURF interest operator presented in Bay et al. 2008); preliminary point pairs are established, looking for a similar score of their descriptor (see Barazzetti et al. 2011); these pairs are then filtered by a RANSAC procedure considering as consistent a projective planar transformation induced by an (unknown) rotation of the support;
3. The transformation is used to resample the input images, removing the effect of unwanted movements of the box;
4. The resampled new images are used in the matching procedure after epipolar resampling (see Pollefeys et al. 1999). The points on the 'master' image are

always selected on the same regular grid (commonly with a 2-pixel spacing) (see Scaioni et al. (2014b));
5. To filter out vegetation or gross matching errors, every matched point on the master grid is compared with its neighbours: if there is a sudden change in the disparity values (corresponding to an equally sudden change in depth in 3D space), the point is labelled as an outlier and discarded;
6. After the matching step, the points extracted are triangulated using the orientation parameters of the reference image pair;
7. A colour-coded map of the displacements measured by the system at the new epoch is automatically generated (see Fig. 6, for instance); and
8. The DSM is stored as ASCII point cloud and as triangulated model (PLY format).

With the settings employed for the test site of Mont de la Saxe (see Sect. 6), the whole process takes about 15 min on a medium-performance server with eight processing cores to produce a DSM with on average 600–700 thousand points (excluding those removed by the filtering step).

5 Surface Point Tracking

Another option that can be exploited with a stereo-vision monitoring system is tracking continuously point features on the landslide slope. While, from a DSM comparison, the volumetric variation along the line of sight (LoS) of the camera is highlighted, with feature tracking real 3D displacement and/or velocity can be computed. Feature tracking allows either following slowly moving points along the image sequence as well as detecting sudden and discontinuous displacements.

The idea is pretty straightforward: instead of comparing the synchronous images of single stereo pairs (that is used to reconstruct the DSM), the same ABM algorithms mentioned in the previous paragraphs can be used to monitor the displacement of subsequent asynchronous images acquired from the same camera. In that case, the perspective differences between homologous points are negligible. In other words, the very same methodology used in traditional *digital image correlation* (DIC) applications (see, for instance, Birgisson et al. 2009) can be used for landslide surface-point tracking (SPT) (see Scaioni et al. (2014b)). Once the displacement of a landslide point is recovered in image space, a back-projection of its former and latter position onto the DSM can reconstruct the movement in the object space. In Fig. 3, the graphical user interface (GUI) of DICe (a proprietary-developed software package of the University of Parma) is shown when elaborating data from the Mont de la Saxe landslide. The same approach can be followed using FBM algorithms as shown in Feng et al. (2012), where FBM is applied with SURF (Bay et al. 2008).

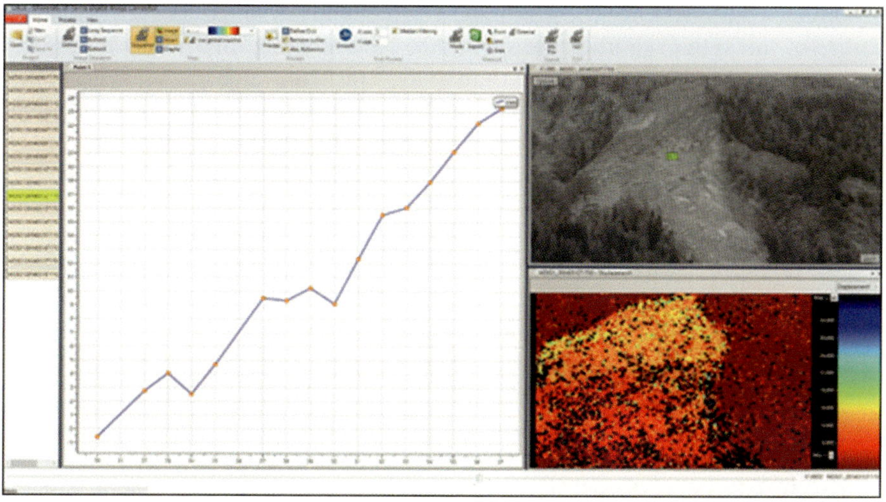

Fig. 3 The GUI of DICe elaborating Mont de la Saxe landslide data

In both cases, the accuracy of tracking depends on the capability of the algorithm to recognize the same area/feature between consecutive frames. If the landslide velocity is low, the radiometric characteristic in the area should vary slowly as well (at least if the lighting conditions do not change too much); consequently, both ABM and FBM algorithms should produce precise, accurate and reliable results. On the other hand, if the landslide velocity is high, the precision of the matching stage is less important, but higher changes in the appearance of the features on the slope surface (e.g. a rolling block) can make the tracking process not reliable or impossible.

As far as FBM tracking is concerned, some of the tests described in Sect. 3.2, conducted by the University of Parma in 2013 on the Mont de la Saxe landslide (see next section), showed that the repeatability of the feature tracking (defined as the ratio of the number of features correctly recognized automatically by the FBM algorithm at some epoch to the total number of features at the first epoch) decreases quite rapidly even for short periods of time (less than 1 month). Moreover, if the reference and current epoch are taken at different times of the day, there is an abrupt fall of the number of features matched by the descriptors. The features, usually, are located by the interest operators on very small elements of the landslide (generally little blocks or group of stones) that create highly contrasted shadows on the landslide surface. Even in a cloudy day, the tests showed that the different position of the sun behind the clouds changes the response of the interest operator as well as the characteristics of the corresponding descriptor. If the image is prefiltered to improve the radiometric distribution of the pixels (e.g. using the *Wallis filter*, Wallis 1974), the issue obviously is more problematic.

On the other hand, if ABM techniques are employed in the SPT algorithm, the ability of the system to track the very same feature along time seems to improve. The different sun position and exposure to light make the algorithm less accurate in determining corresponding positions, but the same feature can be tracked for very long periods, unless considering very different time of the day between the two images or unless the local shape of the slope has changed dramatically. At the same time, false positives are quite frequent since similar pattern of pixels can be found quite easily on the landslide surface: usually to discard a matched position, the results obtained by the ABM algorithm are checked evaluating the similarity between two image areas using some meaningful statistical operator. For instance, many algorithms use the normalized cross-correlation score of the matched areas and consider the feature correctly identified if such value is higher than a predefined threshold. In outdoor applications, such threshold should not be too restrictive since, as already pointed out, different illumination conditions due to weather or sun position can reduce the similarity of corresponding areas.

An efficient improvement to the problem is represented by the use of global or semi-global ABM algorithms (for a brief review, see Szeliski 2010): all these methods imply a sort of regularity and/or continuity of the pixel displacement field between the two images, enforcing a constraint that, in such a case, makes the identification of corresponding areas more robust and provides results physically more correct. In many cases, however, the formulation of the minimization scheme should be modified from that usually presented in the literature. Our research group is currently involved in modifying the DICe workflow with the aim to integrate in the point tracking stage such methodologies as well.

6 Landslide Monitoring: The 'Mont de La Saxe' Case

The landslide of Mont de la Saxe in Courmayeur (AO—Italy, Fig. 4), above the hamlet of the same name, is considered one of the most complex landslides in Italy. With an estimated volume of about 8 million cubic metres (Crosta et al. 2012), the movements threaten the villages of Entreves and La Palud as well as the A5 motorway and the national road SS 26. In the spring 2013, a sudden acceleration forced the local administration to evacuate the area and to close all the accesses to the Ferret Valley. A second, more critical acceleration, was registered during spring 2014, which forced the local administration to close the A5 motorway for some hours.

The landslide is monitored since 2009. A slow continuous movement alternates with sudden accelerations. This complex dynamics drew international attention not only to assess the geological and geotechnical problems, but also to evaluate and compare different state-of-the-art monitoring techniques. In particular, the trend of the phenomenon is now being followed by seven permanent Global Positioning

Fig. 4 The Mont de la Saxe landslide (25th June 2013)

System (GPS) stations, 25 optical prisms monitored every 2 h by a robotic total station, and a ground-based InSAR (GBSAR) system with a cell resolution from 0.5 to 2.2 m in the range 500–1,000 m. A displacement map in the line-of-sight direction is produced every 4 h; the displacement time series of 10 points distributed on the landslide body is recorded. A SAR image is acquired every 7 min. In addition, periodic surveys based on TLS are accomplished to produce updated DSM of the landslide.

The stereo photogrammetric system presented in Sect. 2 was tested to monitor shape changes of the scene, and its first results are presented in this paragraph.

For easier system management and safety reasons, one of the stations was installed near the Fondazione Montagna Sicura (FMS—a no-profit organization that studies risks and safety in high mountain environment) facilities; the second box was installed at about 150 m from the first, at a slightly lower elevation. The location was carefully chosen to obtain a homogenous frame scale.

The distance from the installation site to the object varies between 500 and 650 m. The pose of the camera stations is slightly convergent to ensure maximum overlap. Considering the camera characteristics and the camera relative geometry and assuming an image matching precision of 0.25 pixels, the theoretical accuracy of ground point coordinates ranges from approx. 5.5 to 9 cm along the mean optical axis direction. A GSD of approx. 20 cm and accuracy between ±5 and ±10 cm were foreseen. Given the landslide dynamics, such values are compatible for a comparison on a monthly basis.

As a consistency test to evaluate the level of performance achieved by the system, it was decided to produce and compare, either automatically and manually, a series of 20 DSM's referring to the time interval that goes from 10th July 2013 to 10th September 2013.

Figure 5 shows an example of a DSM produced automatically by the system.

The DSM's were selected considering the best days and time of the day according to two criteria: first, trying to monitor some periods of consecutive days

Fig. 5 The DSM of the slope acquired on 23rd July 2013 at 13:00

evenly distributed on the two-month period and then selecting pairs with the higher number of matched points. Comparing DSM's of consecutive days, the measurement repeatability can be evaluated: on such short periods, the possible landslide displacements can be considered insignificant for the system accuracy level, and the models should not manifest any change. In other words, the differences shown by the comparison should be considered derived by the measurement noise of the system itself. The comparison is presented in Fig. 6, in which colours represent the distances (absolute differences) between two models separated by 4 h. Most of the points are coloured in blue (corresponding to a movement between 2 and 4 cm) with peaks reaching, in some cases (green), approx. 10 cm. It should be noted that usually the maximum discrepancies show up in those areas where the complexity of the object or the presence of shadows makes the matching procedure more troublesome.

Some remarks can be done on the *pointwise* evaluation of the differences (in this case considering the signed distance of more than 600,000 points) and on analysis of their spatial pattern. The data are approximately normally distributed with a mean distance between the two epochs considered in the different comparisons usually not null. A sort of systematic misalignment of both DSM's (likely due to a residual, not modelled box movement) is found quite often. However, the mean

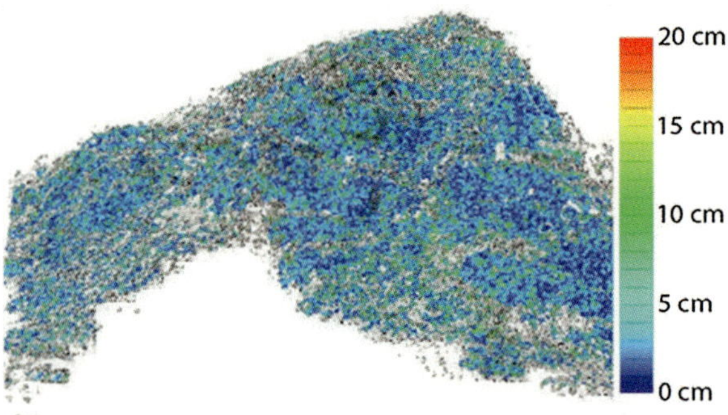

Fig. 6 Repeatability between two 4-h-separated DSM's

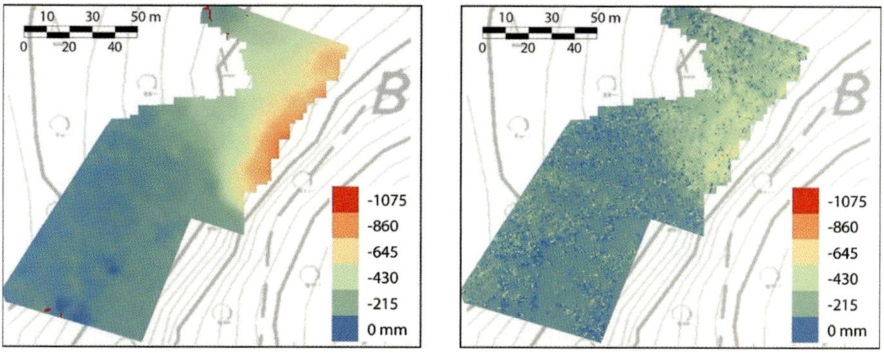

Fig. 7 Side-by-side comparison of displacement maps acquired on 9th October 2013: the InSAR displacement map is shown on the *left* and the photogrammetric displacement map on the *right*

distance between two epochs is always lower than 2 cm and can be considered negligible if compared to the actual system accuracy.

Finally, the data extracted daily in a period of time much longer (approx. 4 months during the summer and fall season) were compared with the displacement maps measured by the GBSAR system. The latter is certified as a displacement measurement system with higher precision, since it may identify movements up to some millimetres.

Figure 7 shows the comparison between two displacement maps, roughly at the middle of the monitoring period (9th October 2013). The GBSAR system monitors a much wider region of the slope, including also areas with overgrown vegetation: in the comparison, such regions have been removed.

The map shows clearly that the photogrammetric system is much noisier than the GBSAR. Nonetheless, the results are similar and both systems achieve the same level of detail for finer and more localized movements.

From a quantitative point of view, a comparison of local displacement values can be more efficient. The GBSAR system, at each epoch, produces a displacement value for a set of 15 'virtual' points on the slope. Some of them are, luckily, placed in the same region monitored by the photogrammetric system and can be used to check the system accuracy. In Fig. 8, the plot of the displacements in the analysed period is shown.

Even if in this comparison the photogrammetric system shows noisier results, the agreement of the two systems is clear. The mean value of the differences is 35 mm, with a standard deviation of 118 mm. However, the maximum observed difference was more than 40 cm high (the negative peak on the 27th September in Fig. 8): the local, pointwise, displacement value can suffer from gross errors and should never be used individually. The system seems to capture the landslide behaviour with accuracy better than expected: by smoothing the observed displacements to remove gross errors and noisy peaks with a short-time running window (i.e. on a weekly basis), quite good results can be achieved.

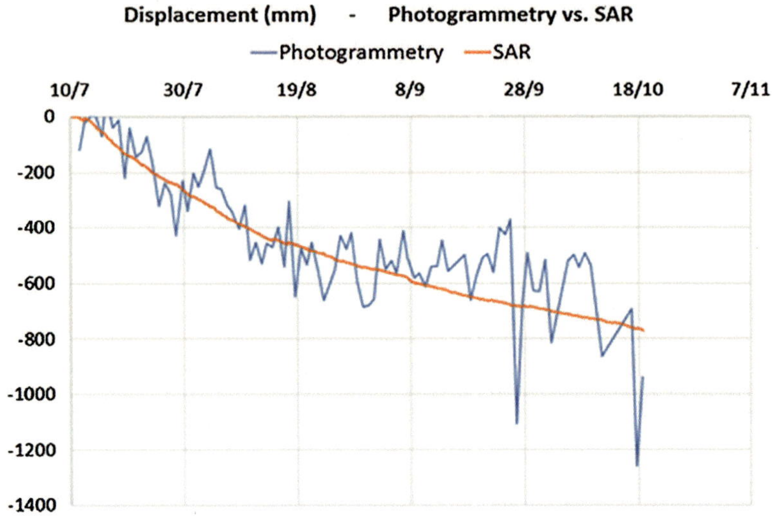

Fig. 8 Photogrammetry versus GBSAR comparison in one of the 'virtual' measurement points on the slope

7 Conclusions

In this chapter, an innovative stereo photogrammetric monitoring system to control the movements of a landslide has been presented. To this day, very few examples can be found in the literature of a similar approach.

State-of-the-art technologies such as ground-based InSAR are too expensive to be diffusely used, unless the risk and potential damages are high; in some cases, access to the landslide to install prisms for theodolite-based monitoring might not be safe. The proposed photogrammetric monitoring system, on the contrary, is very cheap and does not require maintenance of targets or access to the landslide body. Since a single stereo pair is used, a compromise between the GSD (i.e. the system sensitivity to displacement) and the extent of the area that can be controlled must generally be made by selection of camera resolution, base length and focal length. Therefore, in some cases, the system accuracy might be fully appropriated to the task; in some others, like the Mont de la Saxe landslide, it will not provide, with enough reliability, information on landslide movements on an hourly or even daily basis. However, on longer time periods, it is able to accurately evaluate the global and local behaviour, especially on smaller slopes.

The testing stage over one of the most studied landslides of the Alps is still underway, to demonstrate the actual performances of the system in sensing the displacement and in continuous operation over a long period of time.

Acknowledgments This research was supported by the Italian Ministry of University and Research within the project FIRB—Futuro in Ricerca 2010 (No. RBFR10NM3Z).

References

Abellán, A., Oppikofer, T., Jaboyedoff, M., Rosser, N. J., Lim, M., & Lato, M. J. (2014). Terrestrial laser scanning of rock slope instabilities. *Earth Surface Processes and Landforms, 39*, 80–97.

Angeli, M., Pasuto, A., & Silvano, S. (2000). A critical review of landslide monitoring experiences. *Engineering Geology, 55*, 133–147.

Aspert, N., Santa-Cruz, D., & Ebrahimi, T. (2002). MESH: Measuring errors between surfaces using the Hausdorff distance. In *IEEE International Conference in Multimedia and Expo* (ICME) 2002 (Vol. 1, pp. 705–708), Lausanne, Switzerland, August 26–29, 2002.

Barazzetti, L., Forlani, G., Remondino, F., Roncella, R., & Scaioni, M. (2011). Experiences and achievements in automated image sequence orientation for close-range photogrammetric projects. In *Proceedings of International Conference 'Videometrics, Range Imaging, and Applications XI'*, Proceedings of SPIE (Vol. 8085, p. 13), Munich, Germany, May 23–26, 2011. Paper No. 80850F. doi:10.1117/12.890116.

Barazzetti, L., Previtali, M., & Scaioni, M. (2013). Stitching and processing gnomonic projections for close-range photogrammetry. *Photogrammetric Engineering and Remote Sensing, 79*(6), 573–582.

Barbarella, M., & Fiani, M. (2013). Monitoring of large landslides by terrestrial laser scanning techniques: Field data collection and processing. *European Journal of Remote Sensing, 46*, 126–151.

Bay, H., Ess, A., Tuytelaars, T., & Van Gool, L. (2008). Speeded-up robust features (SURF). *Computer Vision and Image Understanding, 110*(3), 346–359.

Bertacchini, E., Capra, A., Castagnetti, C., & Corsini, A. (2011). Atmospheric corrections for topographic monitoring systems in landslides. In *Proceedings FIG Working Week 2011* 'Bridging the Gap between Cultures' (p. 15), Marrakech, Morocco, May 18–22, 2011.

Birgisson, B., Montepara, A., Romeo, E., Roncella, R., Roque, R., & Tebaldi, G. (2009). An optical strain measurement system for asphalt mixtures. *Materials and Structures, 42*(4), 427–441.

Bitelli, G., Dubbini, M., & Zanutta, A. (2004). Terrestrial laser scanning and digital photogrammetry techniques to monitor landslide bodies. *International Archives of The Photogrammetry, Remote Sensing and Spatial Information Sciences, 38*(7B), 246–251.

Brown, D. (1971). Close-range camera calibration. *Photogrammetric Engineering, 37*(8), 855–866.

Cardenal, J., Mata, E., Perez-Garcia, J., Delgado, J., Andez, M., Gonzales, A., et al. (2008). Close range digital photogrammetry techniques applied to landslide monitoring. *International Archives of The Photogrammetry, Remote Sensing and Spatial Information Sciences, 37*(B8), 235–240.

Casson, B., Baratoux, D., Delacourt, C., & Allemand, P. (2003). La Clapière landslide motion observed from aerial differential high resolution DEM. *Engineering Geology, 68*, 123–139.

Cignoni, P., Rocchini, C., & Scopigno, R. (1998). Metro: Measuring error on simplified surfaces. In *Proceedings Computer Graphics Forum* (Vol. 17, No.2, pp. 167–174) June 1198.

Corominas, J., Moya, A., Lloret, J., Gili, A., Angeli, M. G., Pasuto, A., et al. (2000). Measurement of landslide displacements using a wire extensometer. *Engineering Geology, 55*, 149–166.

Crosetto, M., Monserrat, O., Luzi, G., Cuevas-Gonzáles, M., & Devanthéry, N. (2014). Discontinuous GBSAR deformation monitoring. *ISPRS Journal of Photogrammetry and Remote Sensing, 93*, 136–141.

Crosta, G., et al. (2012). Chasing a complete understanding of a rapid moving rock slide: The La Saxe landslide. In *Proceedings of EGU General Assembly* 2012. Vienna, Austria.

Feng, T., Liu, X., Scaioni, M., Lin, X., & Li, R. (2012). Real-time landslide monitoring using close-range stereo image sequences analysis. In *Proceedings of 2012 International Conference on Systems and Informatics (ICSAI 2012)* (pp. 249–253), Yantai, P.R. China, May 19–21, 2012.

Fischler, M., & Bolles, R. (1981). Random sample consensus: A paradigm for model fitting with applications to image analysis and automated cartography. *Communications of the ACM, 24*(6), 381–395.

Forlani, G., & Pinto, L. (1994). Experiences of combined block adjustment with GPS data. *International Archives of Photogrammetry and Remote Sensing, 30*(3/1), 219–226.
Forlani, G., & Pinto, L. (2007) GPS-assisted adjustment of terrestrial blocks. In Proceedings of 5th International Symposium on 'Mobile Mapping Technology (MMT'07)' (pp. 1–7). Padova, Italy.
Fraser, C. S. (1996). Network design. In K. B. Atkinson (Ed.), *Close range photogrammetry and machine vision* (pp. 256–281). Dunbeath, Caithness, Scotland, UK: Whittles Publishing.
Grün, A. (1985). Adaptative least squares correlation: a powerful image matching technique. *South African Journal of Photogrammetry, Remote Sensing and Cartography, 14*, 175–187.
Hartley, R., & Zissermann, A. (2006) *Multiple view geometry in computer vision*. Cambridge: Cambridge University Press.
Intrieri, E., Gigli, G., Casagli, N., & Nadim, F. (2013). Landslide early warning system: Toolbox and general concepts. *Natural Hazards and Earth System Science, 13*, 85–90.
Jaboyedoff, M., Oppikofer, T., Abellán, A., Derron, M. H., Loye, A., Metzger, R., et al. (2012). Use of LIDAR in landslide investigations: A review. *Natural Hazards, 61*, 1–24.
Jacobsen, K. (2004). Direct integrated sensor orientation-pros and cons. *International Archives of the Photogrammetry, Remote sensing and Spatial Information Sciences, 35*(B3), 829–835.
Kraus, K. (2008). *Photogrammetry—geometry from images and laser scans*. Berlin, Germany: Walter de Gruyter.
Luhmann, T., Robson, S., Kyle, S., & Böhm, J. (2013). *Close range photogrammetry: 3D imaging techniques* (p. 702). Germany: Walter De Gruyter Inc.
Monserrat, A., & Crosetto, M. (2008). Deformation measurement using terrestrial laser scanning data and least squares 3D surface matching. *ISPRS Journal of Photogrammetry and Remote Sensing, 63*, 142–154.
Monserrat, O., Crosetto, M., & Luzi, G. (2014). A review of ground-based SAR interferometry for deformation measurement. *ISPRS Journal of Photogrammetry and Remote Sensing, 93*, 40–48.
Mora, P., Baldi, P., Casula, G., Fabris, M., Ghirotti, M., Mazzini, E., et al. (2003). Global Positioning Systems and digital photogrammetry for the monitoring of mass movements: Application to the Ca' di Malta landslide (northern Apennines, Italy). *Engineering Geology, 68*, 103–121.
Motta, M., Gabrieli, F., Corsini, A., Manzi, V., Ronchetti, F., & Cola, S. (2013). Landslide displacement monitoring from multi-temporal terrestrial digital images: Case of the Valoria landslide site. In C. Margottini, P. Canuti & K. Sassa (Eds.), *Landslide science and practice* (Vol. 2, pp. 73–78). Heidelberg, Berlin: Springer.
Pollefeys, M., Koch, R., & Van Gool, L. (1999). A simple and efficient rectification method for general motion. In *Proceedings of '7th IEEE International Conference on Computer Vision (ICCV'99)'* (Vol. 1, pp. 496–501).
Qiao, G., Lu, P., Scaioni, M., Xu, S., Tong, X., Feng, T., et al. (2013). Landslide investigation with remote sensing and sensor network: From susceptibility mapping and scaled-down simulation towards in situ sensor network design. *Remote Sensing, 5*(9), 4319–4346.
Raguse, K., & Heipke, C. (2009). Synchronization of image sequences—a photogrammetric method. *Photogrammetric Engineering and Remote Sensing, 75*(4), 535–546.
Scaioni, M., Roncella, R., & Alba, M. (2013). Change detection analysis in point clouds: Application to rock face monitoring. *Photogrammetric Engineering and Remote Sensing, 78* (5), 441–455.
Scaioni, M., Longoni, L., Melillo, V., & Papini, M. (2014a). Remote sensing for landslide investigations: An overview on recent achievements and perspectives. *Remote Sensing, 6*(10), 9600–9652. doi:10.3390/rs6109600.
Scaioni, M., Feng, T., Lu, P., Qiao, G., Tong, X., Li, R., et al. (2014b). Close-range photogrammetric techniques for deformation measurement: Applications to landslides. In M. Scaioni (Ed.), *Modern technologies for landslide investigation and prediction* (pp. 13–41). Berlin Heidelberg: Springer.
Szeliski, R. (2010). *Computer vision: Algorithms and applications*. Berlin Heidelberg: Springer.

Travelletti, J., Delacourt, C., Allemand, P., Malet, J. P., Schmittbuhl, J., Toussaint., R. et al. (2012). Correlation of multi-temporal ground-based optical images for landslide monitoring: Application, potential and limitations. *ISPRS Journal of Photogrammetry and Remote Sensing, 70*, 39–55.

Wallis, K. F. (1974). Seasonal adjustment and relations between variables. *Journal of the American Statistical Association, 69*(345), 18–31.

Wasowski, J., & Bovenga, F. (2014). Investigating landslides and unstable slopes with satellite multi temporal interferometry: Current issues and future perspectives. *Engineering Geology, 174*, 103–138.

A New Approach Based on Terrestrial Remote-sensing Techniques for Rock Fall Hazard Assessment

Paolo Mazzanti, Alessandro Brunetti and Alberto Bretschneider

Abstract Remote-sensing techniques are changing the way of investigating the Earth and its surface processing. Among these, rock fall from vertical cliffs are very frequent and difficult to be investigated because they frequently occur from inaccessible places. At this regard, terrestrial remote-sensing techniques represent a great opportunity for investigating inaccessible cliffs from a remote position. In this paper, a new approach for the investigation of rock cliff and the prioritization of rock fall hazard based on data collected by remote-sensing techniques has been developed and applied to a real coastal cliff located in the southern part of Italy. By the herein presented approach, data derived from a survey performed by the combination of terrestrial laser scanner, ground-based SAR interferometry and infrared thermography are used in order to identify both predisposing factors (mapping of discontinuities) and state of activity indicators of rock instabilities. Hence, a prioritizations map of the cliff in terms of stability interventions is achieved that can be easily used by local authorities in charge of land management.

Keywords Ground-based SAR interferometry · Terrestrial laser scanning · Infrared thermography · Priority maps · State of activity

P. Mazzanti (✉)
Department of Earth Sciences, 'Sapienza' University of Rome, P.le A. Moro 5, 00185 Rome, Italy
e-mail: paolo.mazzanti@nhazca.com

P. Mazzanti · A. Brunetti
NHAZCA S.r.l., spin-off 'La Sapienza' University of Rome, via Cori snc, 00177 Rome, Italy
e-mail: alessandro.brunetti@nhazca.com

P. Mazzanti
CERI, Research Centre for the Prevention, Prediction and Control of Geological Risks, 'LA Sapienza' University of Rome, P.le A. Moro 5, 00185 Rome, Italy

A. Bretschneider
Department of Geotechnics, Environment, Natural Hazards and Earth Sciences, IFSTTAR, Nantes, France
e-mail: alberto.bretschneider@ifsttar.fr

M. Scaioni (ed.), *Modern Technologies for Landslide Monitoring and Prediction*, Springer Natural Hazards, DOI 10.1007/978-3-662-45931-7_4

1 Introduction

Falls, topples and large collapses from cliffs are among the most risky gravity-induced slope instabilities (Fell 1994; Hunger and Evans 1989). Field mapping of rock discontinuities is the most ordinary approach for the analysis of rock cliffs. This classical geomechanical investigation method allows the main joints to be characterized in terms of dip and dip direction, opening, presence of gouge material, persistency, etc. (see Hatheway 2009 for a comprehensive review). Furthermore, the mechanical features of the rock mass such as the surface roughness, the rock compressive strength and other similar properties can be assessed by suitable field instrumentations. Finally, by an accurate geomechanical analysis, potentially unstable blocks or cliff sectors can be identified and a safety factor for the cliff slope can be estimated.

Nevertheless, high cliffs are very difficult to be accessed and therefore they often cannot be investigated by traditional field methods. To this aim, terrestrial remote-sensing techniques may offer a new opportunity.

In this paper, an innovative method for the susceptibility analysis of rock falls is presented. The method is based on the combination of data derived from *terrestrial laser scanning* (TLS) (see Wehr and Lohr (1999), Fröhlich and Mettenleiter (2004)), *terrestrial infrared thermography* (see Gaussorgues (1994)), and *ground-based SAR interferometry* (GBSAR) surveys (see Bozzano et al. (2010) and Mazzanti (2011)). The innovative method has been tested on a 45-m-high coastal cliff, made up of limestones, located in the Gargano peninsula (southern Italy). Being the cliff inaccessible, the application of remote-sensing techniques is the only solution for the geomechanical characterization of the rock mass (see also Martino and Mazzanti (2014)).

2 Remote-sensing Investigation of Natural Cliffs

Over the last decades, several ground-based remote-sensing techniques have been developed. Techniques based on both passive (i.e. photogrammetry and infrared thermography) and active (i.e. Laser, Radar) sensors are opening new opportunities for the visualization and investigation of natural rock cliffs. In the frame of this research, three different techniques have been used and integrated in order to collect data of a coastal cliff. In the following subsections, a short description of the basic principles of these techniques is presented.

2.1 Terrestrial Laser Scanner

TLS is a topographic technique that allows the collection of 3D point clouds of the surveyed area (see Heritage and Large (2009), Shan and Toth (2009) and Vosselman and Mass (2010), Buckley et al. (2008)). A terrestrial laser scanner consists of an active near-infrared laser, which may operate on the basis of different principles.

The most common operational principle is the measurement of the time-of-flight (TOF), which is the elapsed time between the emission of the laser pulse and the reception after backscattering on the surface of the surveyed object. The laser beam is rotated to scan the portion of the object to be measured. Usually, a set of points (also termed as 'point cloud') are recorded. The direct observations gathered by the instrument consist in the polar coordinates of any measured points. Such coordinates are transformed first into a local cartesian coordinate system (*intrinsic reference system* or IRS) by using elementary mathematical formulas. In a following step, the point cloud can be then translated into an external reference system (*ground reference system* or GRS) by the support of a few *ground control* points (GCP's), whose 3D coordinates have to be independently measured by using geodetic techniques (GNSS or theodolites). The integration of a high-resolution digital camera allows adding true colour information to the point cloud. The point spacing of the point cloud depends on the sensor-to-target distance and on the instrumental characteristics: different values can be set up depending on the wanted point density. In addition, when a laser beam hits the surface of the object at a prefixed angular position, the range is not measured on a single point but it is averaged on the whole footprint of the laser beam, which tends to diverge when travelling out from the instrument.

More than one point cloud may be gathered from different scan positions so that large objects can be fully covered and shadowed zones can be strongly reduced. Furthermore, the point cloud density can be significantly increased, thus reaching a millimetre distance among the points.

By the analysis of the high-resolution point cloud, the main joint systems of the rock masses and their spatial orientation can be detected, thus identifying unstable rock blocks and computing their volumes (see Kemeny et al. (2006), Sturzenegger and Stead (2009)).

TLS is also adopted for the monitoring of displacements (see Scaioni et al. (2013), Oppikofer et al. (2008), Abellán et al. (2010, 2014)), but its accuracy in the coordinates detection (about some centimetres at a distance of 100 m) may not be sufficient for the monitoring of millimetric displacements.

2.2 *Infrared Thermography*

The infrared (IR) thermography consists in the detection of the temperature through the measurement of the emitted IR radiation of a body by passive IR cameras (Vollmer and Müllmann 2010; Budzier and Garlach 2011). As a matter of fact, bodies with a temperature higher than the absolute zero emit IR radiation whose energy depends on the wavelength, according to the Plank's law:

$$E = \frac{h \cdot c}{\lambda}$$

with h = Plank's constant; c = speed of light; λ = wavelength.

The IR sensors can detect such a radiation and convert it into an electric voltage output which is then amplified and converted into a digital impulse, thus allowing the detection of the temperature value. The final output is a 2D false colour digital photo (*thermogram*), where the colour of each pixel depends on the detected temperature. Some thermal cameras are also equipped with a digital optical camera in order to detect both the IR and the visible signals.

2.3 Ground-Based SAR Interferometry

Ground-based interferometric synthetic aperture radar (GBSAR) is a remote-sensing technique for the measurement of displacements (see Mazzanti 2011 and a review in Monserrat et al. 2014)). GBSAR is based on the same operational principles of the satellite SAR interferometry, as described in Massonnet and Feigl (1998), Ferretti et al. (2007). Both ground-based and satellite SAR techniques make use of an active sensor emitting and receiving microwave signals from different positions in the space while moving along a predefined path (an orbit for a satellite or a linear scanner for a GBSAR). Hence, bi-dimensional maps with a range resolution (along the instrumental line-of-sight or LOS) and a cross-range resolution (along the orthogonal direction with respect to the LOS) can be obtained. By the interferometric measurement of the phase difference of the signal backscattered by the investigated objects at different time, the displacement along the LOS of each pixel and displacement maps can be achieved. Furthermore, the displacement time history for each pixel can also be inferred by multi-stack analyses. The displacement accuracy ranges from few tenths of millimetres to few millimetres depending on the sensing distance and on the reflectivity properties of the investigated surface. The instrumental sampling frequency depends on the adopted sensor (see Monserrat et al. 2014), but generally is in the order of a few minutes. The pixel size ranges from few decimetres to some metres depending on the sensor-to-target distance.

3 Priority Mapping of Natural Cliffs

In order to characterize the overall cliff stability and to define the intervention stabilization priorities (i.e. classifying different cliff sectors on the basis of the most 'urgent' requirement of stabilization), a new method fully based on collected remote-sensing data was developed. This new method takes into account both the preconditioning factors of the cliff and the *state of activity* (SA), thus allowing a comprehensive view of the instability problem. First, an expected rock fall magnitude (ERFAM) class index is assigned to different cliff portions, based on geomechanical investigations performed using high-resolution DTM derived from TLS point cloud coupled with geological data achieved by both literature documents and by in situ surveys. Then, SA value for each cliff portion is attributed using different information from GBSAR and IR data by looking at their displacements and anomalous thermal behaviour (e.g. non-homogeneous spatial and temporal temperature variation in specific sectors of the cliff). Finally, by a suitable combination of ERFAM and SA mapping, the priority mapping of the cliff is achieved. In what follows an extensive description of the above-mentioned analysis is presented. This approach has been developed and firstly applied in the frame of a project for the rock fall hazard assessment of the Monte Pucci coastal cliff in the Gargano Peninsula (Southern-Eastern Italy).

3.1 Geological Setting of the Monte Pucci Coastal Cliff

The detailed geological setting of the study area is treated in Martino and Mazzanti (2014).

The Monte Pucci relief is located in the northern sector of the Gargano Promontory in the northern part of the Apulia region (Southern-Eastern Italy) (Fig. 1).

The Gargano promontory is constituted by carbonate successions deposited during the Upper Cretaceous and slightly folded during the Neogene compressive tectonic phase (Bertotti et al. 1999), forming a large anticlinal structure. This structure is interested by several faults characterized by variable orientation (Gambini and Tozzi 1996; Chilovi et al. 2000). In the north-eastern part of the Gargano promontory (Fig. 2), the basin deposits (formations of: Maiolica, Scisti a Fucoidi and Scaglia) are overlaid, with an on-lap unconformity contact, by the

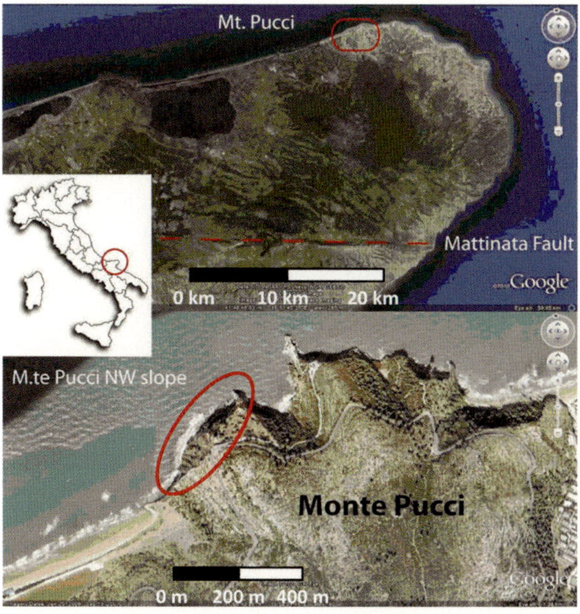

Fig. 1 Geographical location of the Gargano Promontory and of 'Monte Pucci'

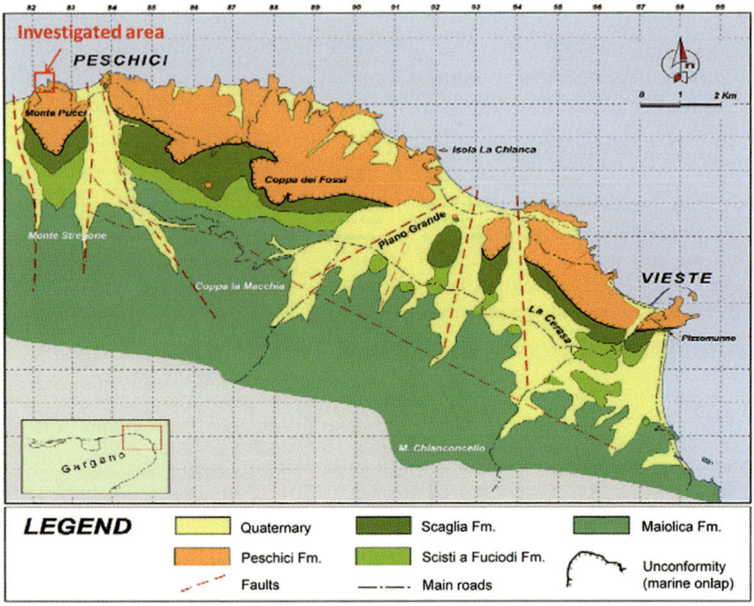

Fig. 2 Geological sketch of the north-eastern sector of the Gargano Promontory (modified from Morsilli 1998) in *red* Monte Pucci is identified

Calcari di Peschici limestones (Speranza and Kissel 1993; Bosellini et al. 1993, 1999). The Monte Pucci reflects this geological setting: the Peschici limestones constitute the top layer of the relief, while the Scaglia formation is at the basis and outcrops only in the lower portion of vertical coastal cliffs.

The north-western cliff of the Monte Pucci relief, investigated in this chapter, is a 45-m-high vertical cliff directly on the sea. The geological limit between the Scaglia and the Calcari di Peschici formations is approx. at 20 m a.s.l. The sub-vertical profile of the cliff is the result of collapses and instabilities that are controlled by two main factors: (a) the marine erosion due to wave action at the toe and (b) the karstification processes acting on the limestone rocks. Karst processes act wide-spread in the region, creating a network of caves and galleries all along the coast. When the roof of these natural caves collapses, relict pillars emerge from the sea.

3.2 Remote-sensing Surveys

A TLS survey was carried out in order to achieve a detailed true colour 3D point cloud of the Monte Pucci cliff. Scanning by a Riegl VZ400 (www.riegl.com) instrument was performed from two different positions in order to reduce the shadowed zones and to obtain a sub-centimetre accuracy of the detected points. The max distance between the instrument and the farthest collected points was about

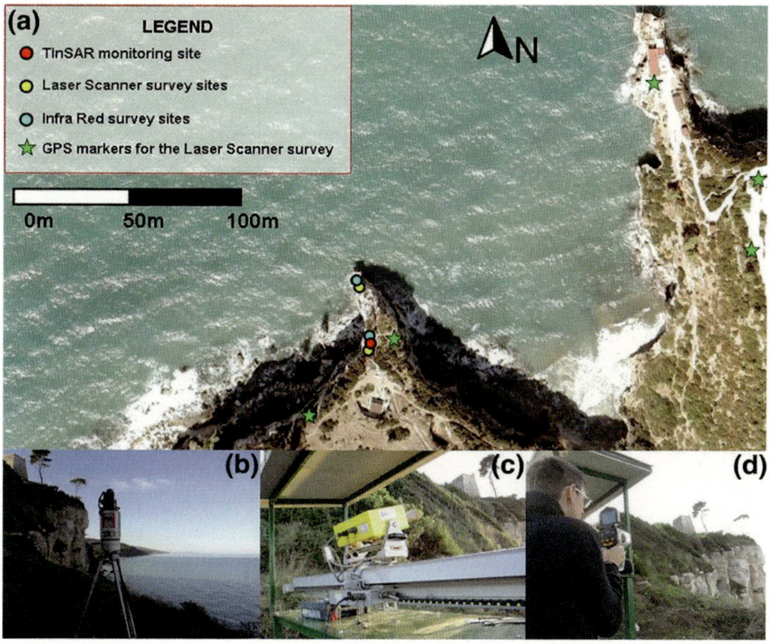

Fig. 3 Plan view of the sensing sites and GPS stations. **a** Photos of the instruments' installations: Riegl VZ400 laser scanner; **b** IDS IBIS-L GBSAR; **c** NEC TH7700 thermocamera; **d** All subfigures have been modified from Martino and Mazzanti (2014)

150 m. Some targets were applied all around the monitored area in order to align the scans from the two stations and for georeferencing into UTM-WGS84 coordinate system (Fig. 3a). Furthermore, by a Nikon D700 digital camera integrated with the VZ400 sensor (Fig. 3b), the RGB colour of the detected points was collected. Hence, a true colour point cloud of the cliff made up of about 45 M points with a mean point density of about 1 point/cm was obtained.

IR thermographic surveys were carried out during 2 days in order to detect thermal images of the cliff surface. Data were collected by a NEC TH7700 camera (Fig. 3d), characterized by a thermal resolution of 0.1 °C, a geometric resolution of 320 × 240 and a thermal sensing range between −20 and +250 °C.

Eleven thermal images were collected from two different stations (see Fig. 3a) in order to obtain a wide coverage of the overall cliff. Moreover, thermograms were gathered at different hours in order to investigate the differential thermal variations of the cliff over time. Hence, both absolute temperature and thermal discrepancies on time were obtained. Furthermore, thermograms collected from different positions at the same time were merged, georeferenced and overlapped to the optical image (Fig. 4).

In order to infer the displacement field of the cliff, a two-day-long continuous GBSAR monitoring survey with a sampling rate of about 5 min was performed, thus collecting a total amount of 442 radar maps. The GBSAR equipment IBIS-L

Fig. 4 Mosaic of collected thermograms overlapping the optical image of the cliff (modified from Martino and Mazzanti 2014)

(IDS S.p.A., Italy, www.idscorporation.com) was housed on a suitably realized iron basement and then installed in front of the investigated rock cliff at a distance ranging from 50 to 150 m (Fig. 3c). The collected SAR images were characterized by a mean range and cross-range resolution of about 0.5 × 0.5 m.

Preliminary georeferencing of the SAR images was performed by identifying common zones on the picture; then, a more accurate georeferencing was performed by the rigorous combination of TLS and GBSAR data. With this aim, radar *corner reflectors* and laser scanning targets were coupled and installed at eight different positions on the cliff by rock climbers. This solution allowed the identification of laser and radar common points to be used for co-registration and georeferencing. Then, through an algorithm specifically developed for this purpose, GBSAR displacement maps were overlapped onto the 3D point cloud derived from TLS surveying, thus obtaining a 3D GBSAR displacement map (Fig. 5).

Conventional and advanced analyses by the 'PtoP' technique, exhaustively described in Mazzanti and Brunetti (2010), were performed on the collected SAR images over the whole investigated portion of the cliff. 'PtoP' is an innovative technique for the analysis of GBSAR data, specifically developed by NHAZCA S.r.l. for the investigation of rock cliffs. This technique allows the measurement of differential displacements between neighbouring pixels in the SAR map, thus assuming, from time to time, reference points in different sectors of the scenario. Through such approach, atmospheric screens affecting data quality are strongly mitigated, thus allowing the measurement of sub-millimetre displacements.

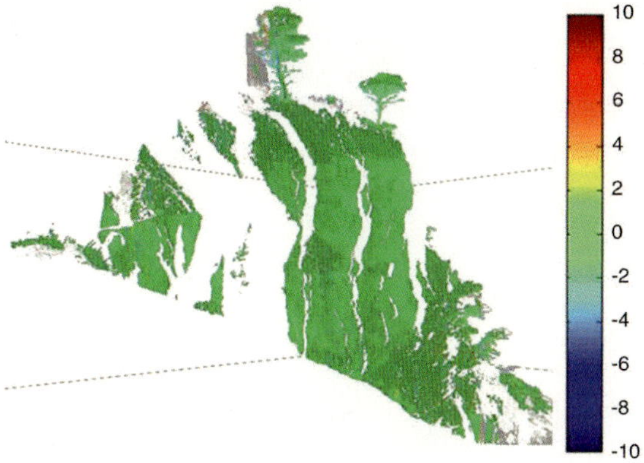

Fig. 5 Projection of the displacement GBSAR map on the 3D TLS-derived point cloud

Hence, pixels affected by irreversible displacement (up to 1.2 mm) and pixel characterized by daily cyclic displacements (between 0.5 and 1 mm) were identified (Fig. 6).

3.3 ERFAM Analysis

In order to identify classes characterized by different ERFAM, both geological and simplified geomechanical analyses were applied.

First of all, both automatic and manual analyses of the discontinuities on the cliff were performed on the TLS point cloud by using Split-FX and AdHoc3D software packages.

The automatic analysis was based on the identification of a sufficient number of points lying on the same surface, thus allowing the automatic measurement of its orientation in the space. Such an approach allowed detailed analyses of the whole outcropping cliff (by analysing a huge amount of surfaces in few minutes), and it was not influenced by the subjectivity of the operator. However, some weak aspects must be accounted when performing an automatic analysis of the geomechanical setting, such as the difficulties to detect not exposed surfaces like the limestone stratification that in our case is parallel to both TLS stations. Furthermore, several shadowed zones are present in the collected point cloud of Monte Pucci cliff, mainly due to logistic concerns during TLS surveying. Specifically, data were collected only by the eastern side of the cliff; hence, the rock joints with a nearly West dip direction could not be identified.

On the other hand, manual analyses of the 3D point cloud are more time-consuming and subjective but they may avoid some of the aforementioned limitations.

Figure 7a shows the final stereoplots derived from the combination of automatic and manual analysis of TLS data for the Calcari di Peschici and Scaglia formations, respectively. As evident, the plots are quite similar to those derived from traditional field surveys (Fig. 7b) collected by 10 geomechanical stations (8 in the surroundings

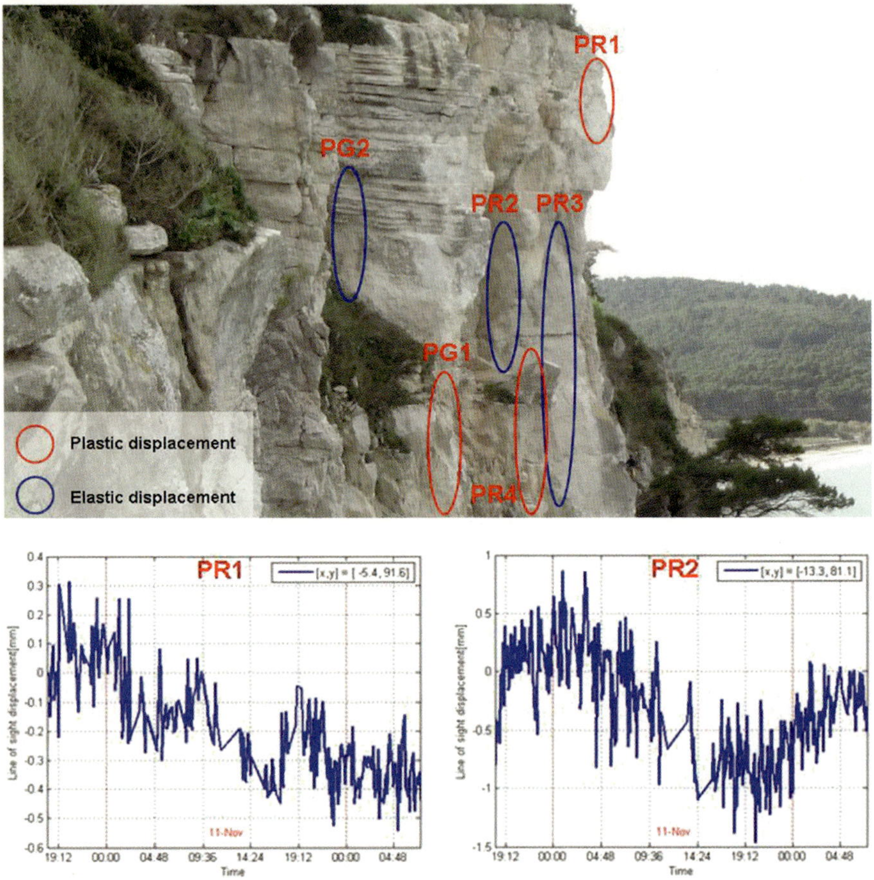

Fig. 6 Identification of pixel affected by plastic and elastic displacements on the cliff image (data by interferometric survey). An example of the two typologies of movements is reported

of the cliff and 2 performed directly on the cliff by two rock climbers). This result confirmed the reliability of the simplified geomechanical analysis derived from TLS data, as already proved by other authors in the literature (Longoni et al. 2012; Riquelme et al. 2014).

Two main joint sets were identified in the Scaglia formation, and four main joint sets were detected in the Calcari di Peschici formation. Detailed analyses of the joint spacing and persistence were performed on the TLS data. In Table 1, the main joint sets and their average spacing values are reported.

All the available data, both from manual and TLS surveys, were used for the computation of the *Jv* parameter (see Palmstrom 1982, 1985, 1986, 1996; Sen and Eissa 1992), thus providing information about the number of joints per m^3. *Jv* values of 3.21 and 9.67 were estimated for the Calcari di Peschici and the Scaglia formations, respectively.

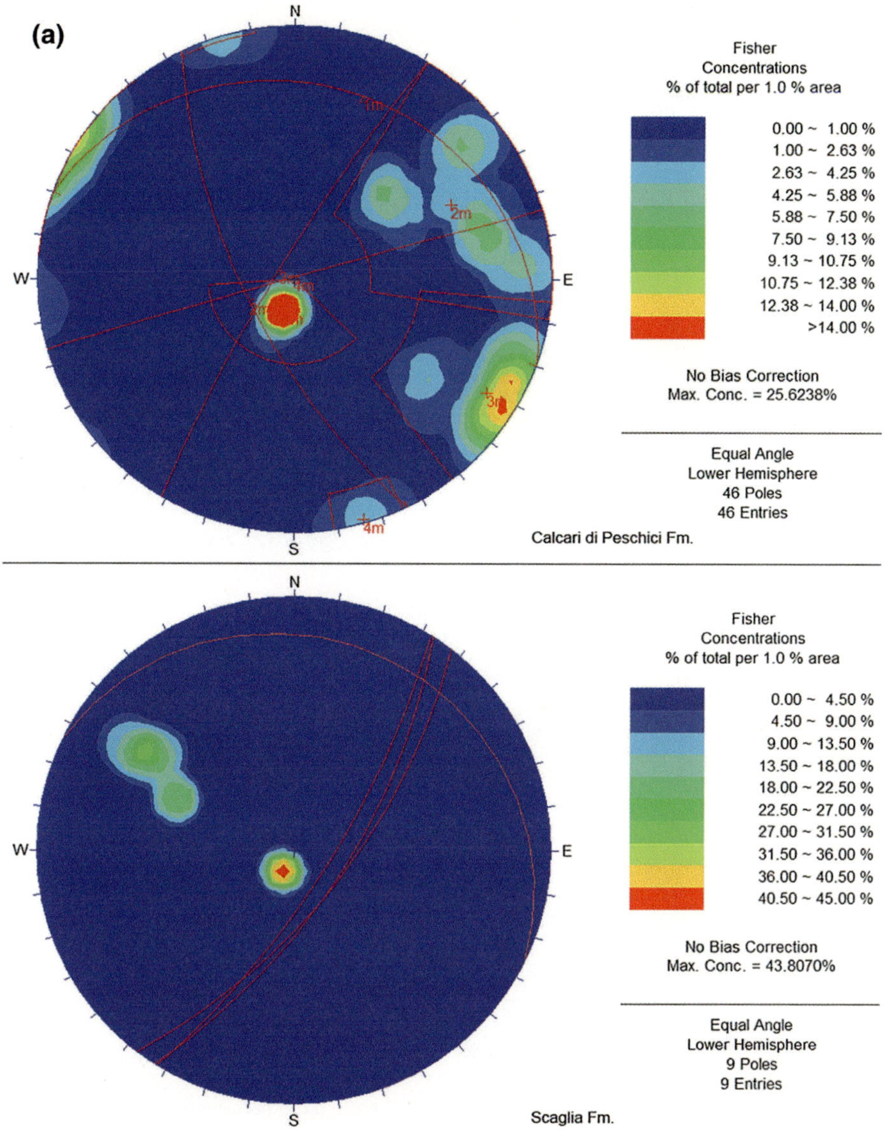

Fig. 7 **a** Stereoplots of the Calcari di Peschici formation (*above*) and of the Scaglia formation (*below*) resulting from the analysis of TLS data (modified from Martino and Mazzanti 2014). **b** Stereoplots of the Calcari di Peschici formation (*above*) and Scaglia formation (*below*) resulting from standard field geomechanical surveys (modified from Martino and Mazzanti 2014)

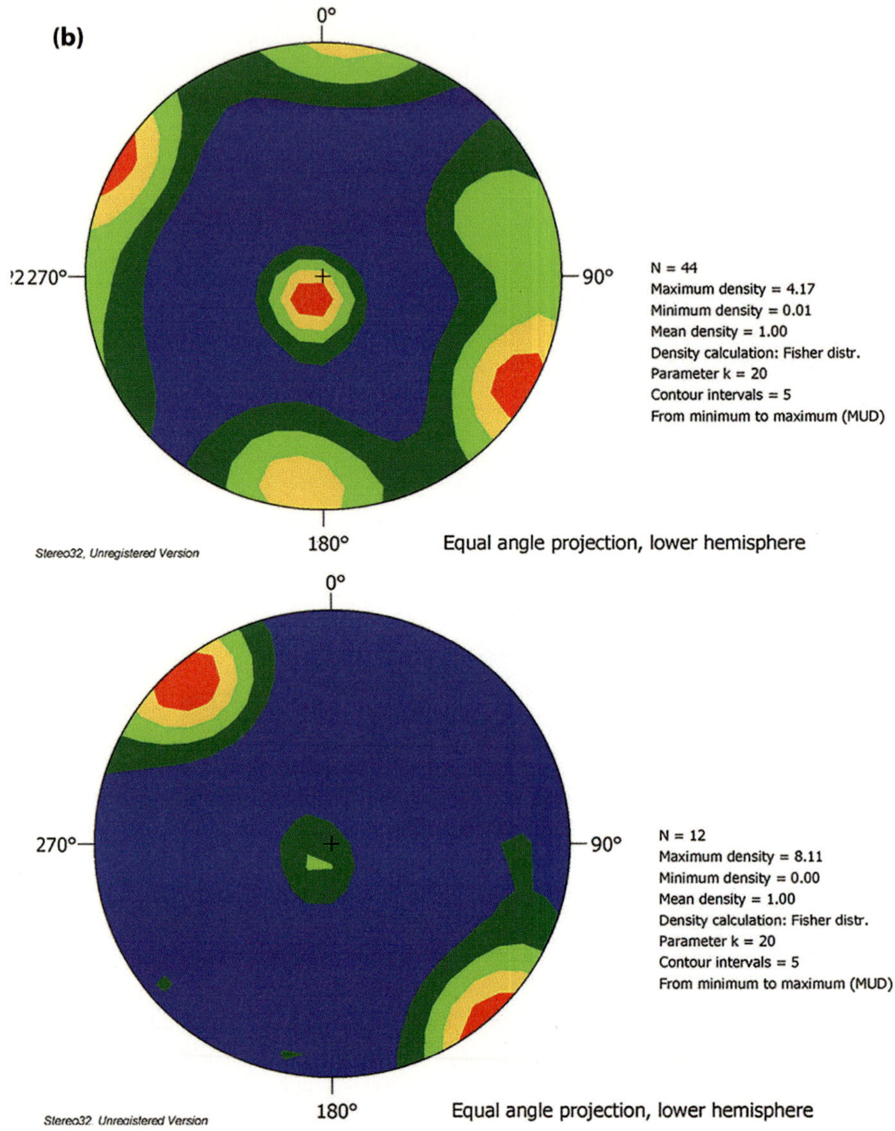

Fig. 7 (continued)

On the basis of these results, a zonation of the cliff has been made by considering 4 ERFAM classes that are described below (Fig. 8):

- ERFAM I: This zone is included in the Calcari di Peschici formation and occupies the eastern cliff sector. The stratification and a medium angle dipping set are the most important joint sets. The intersection of these joint sets with the topographic surface separates wedge blocks with a mean volume of 20 m^3;

Table 1 Main joint sets resulting from automatic TLS data analysis. The average spacing is a mean value referred to all the joint sets. Spacing values are expressed in metres

Formation	Set 1	sp.	Set 2	sp.	Set 3	sp.	Set 4	sp.
Calcari di Peschici formation	20/15	0.64	253/75	0.39	300/82	0.59	350/87	0.54
Scaglia formation	28/10	0.21	120/82	0.67				

Fig. 8 Geomechanical zoning of the Monte Pucci cliff. *Roman numbers* identify different ERFAM classes. For the ERFAM value of each class, please refer to Table 2

- ERFAM II: This zone is included in the Calcari di Peschici formation and located in the Western cliff sector. The most important joint sets are the stratification and some sub-vertical conjugate sets, which are responsible for the identification of vertical prismatic blocks. In this sector of the cliff, the medium angle fractures are absent. Stratification spacing values are about 0.5 m, and the rock blocks volumes range from 1 to 5 m^3;
- ERFAM III: This zone is topographically below the sector II, and it is characterized by the same lithology and the same joint pattern. Nevertheless, the larger stratification spacing (7 ÷ 10 m) isolates rock blocks with a volume of about 15 m^3; and
- ERFAM IV: This zone is comprised in the Scaglia formation. The joint pattern is constituted by the dense stratification, characterized by a middle spacing of about 0.2 m and the sub-vertical joint set corresponding to the slope. Due to this setting, no rock blocks have been detected here.

Table 2 Assessing ERFAM (expected rock fall magnitude) classes by the cliff geomechanical settings

ERFAM classes	Sector I	Sector II	Sector III	Sector IV
Low				•
Medium		•		
High	•		•	

For each class, an ERFAM value has been derived from low to high values (Table 2).

3.4 State of Activity Analysis

Displacement data derived from GBSAR measurement and TIR features of the cliff were used as indicators for the SA estimation of the overall rock cliff.

Specifically, not reversible and cyclic displacements larger than 0.5 mm detected by GBSAR monitoring have been considered as 'displacement anomalies' (Fig. 6) and thus considered as indicators of a higher activity in the related sectors. Then, cliff sectors or single blocks characterized by these anomalies have been mapped as showed in Fig. 9.

Fig. 9 Mapping of the main thermal anomalies detected on the cliff image

Furthermore, it has been assumed that thermal anomalies identify sectors characterized by higher rock fall susceptibility. Specifically, negative thermal anomalies have been associated with the presence of water, positive thermal anomalies to the occurrence of recent rock fall events and differential anomalies to a higher thermoclastic action.

Then, the main thermal anomalies have been mapped as shown in Fig. 9. Specifically, portions of the cliff characterized by extreme temperatures (high and low) have been identified and mapped by using a single thermal image, while the portions characterized by higher temperature gradients (derived from multi-temporal analyses) have been identified by differential thermal images.

3.5 Priority Classes

Information about the predisposition to rock fall (derived from the ERFAM analysis) and about the dynamics of the blocks (derived from the SA analysis) have been integrated in order to define priority maps. Five priority classes have been defined on the basis of the following criteria:

- Priority 1: high ERFAM and at least two independent states of activity indexes (one from GBSAR displacement analysis and the other from TIR analysis);
- Priority 2: high ERFAM and one SA index (GBSAR displacement or TIR), or medium ERFAM and at least two independent SA indexes (GBSAR and TIR);
- Priority 3: high ERFAM and no SA index, or medium ERFAM and one SA index (GBSAR or TIR);
- Priority 4: medium ERFAM and no SA indexes, or low ERFAM and two SA indexes (GBSAR and TIR); and
- Priority 5: medium ERFAM and no SA index, or low ERFAM and one SA index (GBSAR or TIR).

It is worth to note that the priority number (from 1 to 5) identifies a decreasing priority of intervention. In other words, sectors characterized by Priority 1 are considered the most risky and therefore the ones that require stabilization plans.

By applying the above suggested method, the investigated Monte Pucci cliff has been classified as shown in Fig. 10. Results show that there are two sectors in the Calcari di Peschici formation characterized by the highest value of priority and that in general, the lower portion of this formation is characterized by higher priority respect to the upper one.

Fig. 10 Priority mapping of the cliff for the susceptibility to rock fall assessment

4 Discussion and Conclusions

A new approach for the stability characterization of not accessible rock cliffs aiming at prioritizing intervention strategies has been proposed and tested over a coastal rock cliff in the Gargano Promontory (south-eastern Italy). The method is based on the combined use of terrestrial remote-sensing techniques such as ground-based InSAR (GBSAR), TLS and thermal infrared (TIR) analysis. All these techniques share the ability to investigate rock cliffs from a remote position; hence, the herein suggested method can be considered as a fully remote investigation approach, except for the need of positioning a few targets for registration and georeferencing of different data sets.

On the basis of the joint pattern features and on the size of the expected rock falls, the cliff has been divided into three main classes according to ERFAM parameter. However, ERFAM characterization must be considered only the starting point for the suggested approach. As a matter of fact, ERFAM characterization is a 'static' evaluation of the cliff condition that it is only based on some predisposing factors potentially leading to rock fall events. Indeed, in the authors' opinion, the 'SA' of the overall cliff cannot be neglected in order to perform a comprehensive prioritization of the rock cliffs stability. Rather the SA is the indicator of the 'active' behaviour of specific portions of the cliff. For example, thermal anomalies are clearly related to the sunstroke and the consequent thermoclastic actions. Moreover, cyclic displacements (observed by GBSAR) is the more direct and evident measure of the actual activity of any blocks. However, it must be pointed out that the mechanical characterization of the rock mass in terms of response to temperature variation and lower daily cyclic displacement is well beyond the author intention and additional studies are required in order to further validate these results.

Notwithstanding, it has been demonstrated that data derived from TIR and GBSAR surveys can be very useful in supporting conventional geomechanical analysis in prioritizing the areas susceptible to rock fall. By such approach, stability analyses can also be performed, thus allowing the identification of the most hazardous sectors of the cliff, as treated in detail in Martino and Mazzanti (2014).

Nevertheless, it is worth to note that while GBSAR and TLS have been extensively used for similar purposes (see Arosio et al. 2009; Mazzanti and Brunetti 2010), the application of TIR for rock mass analyses is new and challenging (see, e.g. Squarzoni 2008; Baroň 2012). Hence, several approximations have been herein assumed for the analysis of the SA based on thermal data. Therefore, further studies on this matter are necessary for a more comprehensive analysis. However, notwithstanding the approximations, we consider the achieved results reliable especially considering that TIR-derived information are supported by GBSAR and TLS data.

Hence, the approach here proposed can be very useful for engineering geologists and geotechnical engineers involved with local authorities in risk assessment projects and stabilization works design. Furthermore, by using only remote-sensing techniques, the risk related to the investigation of not accessible rock cliffs (conventionally performed by rock climbers) can be significantly reduced.

Future researches will be addressed to the analysis of the kinematic compatibility of the sectors of the slope characterized by high-priority classes and to additional calibrations and validations with results derived by traditional field surveys. Applications to other rock cliffs are also devised.

References

Abellán, A., Calvet, J., Vilaplana, J. M., & Blanchard, J. (2010). Detection and spatial prediction of rock falls by means of terrestrial laser scanner monitoring. *Geomorphology, 119*, 162–171.

Abellán, A., Oppikofer, T., Jaboyedoff, M., Rosser, N. J., Lim, M., & Lato, M. J. (2014). Terrestrial laser scanning of rock slope instabilities. *Earth Surface Processes and Landforms, 39*(1), 80–97.

Arosio, D., Longoni, L., Papini, M., Scaioni, M., Zanzi, L., & Alba, M. I. (2009). Towards rockfall forecasting through observing deformations and listening to microseismic emissions. *Natural Hazards and Earth System Science, 9*(4), 1119–1131.

Baroň, I., Bečkovský, D., & Míča, L. (2012). Application of infrared thermography for mapping open fractures in deep-seated rockslides and unstable cliffs. *Landslides, 11*, 15–27.

Bertotti, G., Casolari, E., & Ricotti, V. (1999). The Gargano Promontory: A neogene contractional belt within the Adriatic plate. *Terra Nova, 11*, 168–173.

Bosellini, A., Neri, C., & Lucani, V. (1993). Platform margin collapses and sequence stratigraphic organization of carbonate slopes: Cretaceous-Eocene, Gargano Promontory. *Terra Nova, 5*, 282–297.

Bosellini, A., Morsilli, M., & Neri, C. (1999). Long-term event stratigraphy of the Apulia Platform margin: Upper Jurassic to Eocene, Gargano, Southern Italy. *Journal of Sedimentary Research, 69*, 1241–1252.

Bozzano, F., Mazzanti, P., Prestininzi, A., & Scarascia Mugnozza, G. (2010). Research and development of advanced technologies for landslide hazard analysis in Italy. *Landslides, 7*(3), 381–385. doi:10.1007/s10346-010-0208-x.

Buckley, S. J., Howell, J. A., Enge, H. D., & Kurz, T. H. (2008). Terrestrial laser scanning in geology: Data acquisition, processing and accuracy considerations. *Journal of the Geological Society London, 165*, 625–638.

Budzier, H., & Garlach, G. (2011). *Thermal infrared sensors, Theory, Optimization and Practice* (p. 302). Chichester, UK: Wiley.

Chilovi, C., De Feyter, A. J., & Pompucci, A. (2000). Wrench zone reactivation in the Adriatic Block: The example of the Mattinata fault system (SE Italy). *Bollettino della Società Geologica Italiana, 119*, 3–8.

Fell, R. (1994). Landslide risk assessment and acceptable risk. *Canadian Geotechnical Journal, 31*, 261–272.

Ferretti, A., Monti Guarnieri, A., Prati, C., Rocca, F., & Massonet, D. (2007). *InSAR principles: Guideline for SAR interferometry processing and interpretation* (Vol. TM-19). Noordwijk, The Netherlands: ESA Publication.

Fröhlich, C., & Mettenleiter, M. (2004). Terrestrial laser scanning—New perspectives in 3D surveying. *International Archives of the Photogrammetry, Remote Sensing and Spatial Information Sciences, 26*(8), W2.

Gambini, R., & Tozzi, M. (1996). Tertiary geodynamic evolution of the Southern Adria microplate. *Terra Nova, 8*, 593–602.

Gaussorgues, G. (1994). *Infrared thermography. Microwave technology* (Vol. 5). London: Chapman & Hall.

Hatheway, H. W. (2009). The complete ISRM suggested methods for rock characterization, testing and monitoring. 1974–2006. *Environmental and Engineering Geoscience, 15*(1), 47–48. doi:10.2113/gseegeosci.15.1.47.

Heritage, G. L., & Large, A. R. G. (2009). *Laser scanning for the environmental sciences* (p. 302). Chichester, UK: Wiley.

Hungr, O., & Evans, S. G. (1989). Engineering aspects of rockfall hazard in Canada. *Geological Survey of Canada, Open File, 2061*, 102.

Kemeny, J., Turner, K., & Norton, B. (2006). LIDAR for rock mass characterization: Hardware, software, accuracy and best-practices. In F. Tonon & J. Kottenstette (Eds.), *Laser and photogrammetric methods for rock face characterization* (pp. 49–62). Alexandria, Egypt: ARMA.

Longoni, L., Arosio, D., Scaioni, M., Papini, M., Zanzi, L., Roncella, R., & Brambilla, D. (2012). Surface and subsurface non-invasive investigations to improve the characterization of a fractured rock mass. *Journal of Geophysics and Engineering, 9*, 461–472.

Martino, S., & Mazzanti, P. (2014). Integrating geomechanical surveys and remote sensing for sea cliff slope stability analysis: The Mt. Pucci case study (Italy). *Natural Hazards Earth System Science, 14*, 831–848. doi:10.5194/nhess-14-831-2014.

Massonnet, D., & Feigl, K. L. (1998). Radar interferometry and its application to changes in the Earth's surface. *Reviews of Geophysics, 36*, 441–500.

Mazzanti, P. (2011). Displacement monitoring by terrestrial SAR interferometry for geotechnical purposes. *Geotechnical instrumentation news, 29*(2), 25–28.

Mazzanti, P. & Brunetti, A. (2010). Assessing rock fall susceptibility by terrestrial SAR interferometry. In *Proceedings of the 'Mountain Risks International Conference'* (pp. 109–114), Firenze, Italy, November 24–26, 2010.

Monserrat, O., Crosetto, M., & Luzi, G. (2014). A review of ground-based SAR interferometry for deformation measurement. *ISPRS Journal of Photogrammetry and Remote Sensing, 93*, 40–48.

Morsilli, M. (1998). Stratigrafia e sedimentologia del margine della Piattaforma Apula nel Gargano (Giurassico superiore-Cretaceo inferiore) (p. 203). PhD dissertation. Italy: Università di Bologna (in Italian).

Oppikofer, T., Jaboyedoff, M., & Keusen, H. R. (2008). Collapse at the eastern Eiger flank in the Swiss Alps. *Nature Geoscience, 1*(8), 531–535.

Palmstrom, A. (1982). The volumetric joint count—A useful and simple measure of the degree of jointing. In *Proceedings of the 4th International Congress IAEG, New Delhi, India* (pp. 221–228).

Palmstrom, A. (1985). Application of the volumetric joint count as a measure of rock mass jointing. In *Proceedings of the International Symposium on 'Fundamentals of Rock Joints', Bjorkliden, Sweden* (pp. 103–110).

Palmstrom, A. (1986). A general practical method for identification of rock masses to be applied in evaluation of rock mass stability conditions and TBM boring progress. *In Proceedings of the Conference on 'Fjellsprengingsteknikk, Bergmekanikk, Geoteknikk', Oslo, Norway, paper No. 31* (pp. 1–31).

Palmstrom, A. (1996). RMi—A system for characterizing rock mass strength for use in rock engineering. *Journal of Rock Mechanics and Tunneling Technology, 1*(2), 69–108.

Riquelme, A., Abellán, A., Tomás, R., & Jaboyedoff, M. (2014). A new approach for semi-automatic rock mass joints recognition from 3D point clouds. *Computers & Geosciences, 68*, 38–52.

Scaioni, M., Roncella, R., & Alba, M. I. (2013). Change detection and deformation analysis in point clouds: Application to rock face monitoring. *Photogrammetric Engineering & Remote Sensing, 79*(5), 441–456.

Sen, Z., & Eissa, E. A. (1992). Rock quality charts for long—Normally distributed block size. *International Journal of Rock Mechanics, Mining Sciences and Geomechanics, 29*(1), 1–12.

Shan, J., & Toth, C. K. (2009). *Topographic laser scanning and ranging. Principles and processing*. Boca Raton, FL, USA: Taylor & Francis Group.

Speranza, F., & Kissel, C. (1993). First paleomagnetism of Eocene rocks from Gargano: Widespread overprint or non rotation? *Geophysical Research Letters, 20*, 2627–2630.

Squarzoni, C., Calgaro, A., Teza, G., Acosta, C. A. T., Pernito, M. A. & Bucceri, N. (2008). Terrestrial laser scanning and infrared thermography in rock fall prone slope analysis. *Geophysical Research Abstracts* 2008, Vol. 10, abstract No. EGU2008-A-09254.

Sturzenegger, M., & Stead, D. (2009). Quantifying discontinuity orientation and persistence on high mountain rock slopes and large landslides using terrestrial remote sensing techniques. *Natural Hazards Earth System Science, 9*, 267–287.

Vollmer, M., & Müllmann, K. P. (2010). *Infrared thermal imaging. Fundamental research and applications* (p. 593). Weinheim, Germany: Wiley-VCH Verlag.

Vosselman, G., & Maas, H. G. (2010). *Airborne and terrestrial laser scanning*. Boca Raton, FL, USA: Taylor & Francis Group.

Wehr, A., & Lohr, U. (1999). Airborne laser scanning—An introduction and overview. *ISPRS Journal of Photogrammetry and Remote Sensing, 54*, 68–82.

Multi-temporal Terrestrial Laser Scanning Survey of a Landslide

Maurizio Barbarella, Margherita Fiani and Andrea Lugli

Abstract Terrestrial laser scanning (TLS) has proven to be a very effective technique for landslides monitoring, even if some critical issues exist for providing highly reliable results. This chapter presents the methodology adopted in performing four surveys, carried out over three years on a large slump landslide in order to get effectively comparable data. The first problem concerns the setting up of the reference system, which has been realized by means of global navigation satellite system permanent stations ETRF00 datum. This solution was able to maximize the stability over time even at the expense of a slightly lower precision, which was, however, in the order of 1–2 cm with data recorded during the whole duration of TLS survey. An assessment of geo-referencing accuracy was carried out with respect to the only stable artifact present in the landslide area. This check pointed out that in the central part of the point cloud the repeatability between different surveys was slightly greater than 5 cm. To ensure the quality of the obtained multitemporal digital terrain models (DTM's) over the entire region of interest, the choice of the interpolation algorithm has been performed and verified with a cross-validation method on the basis of a sample extracted from the data set. To detect the kinematics of the landslide in its several parts, both the DTM's and profiles have been used, which have proven to be particularly useful for the interpretation of details. After the localization of various landslide bodies (keeping into account slope and aspect maps derived from the DTM), the evaluation of the volumes mobilized over time has been carried out by

M. Barbarella (✉) · A. Lugli
Dipartimento di Ingegneria Civile, Chimica, Ambientale e dei Materiali,
University of Bologna, Viale Risorgimento 2, 40136 Bologna, Italy
e-mail: maurizio.barbarella@unibo.it

A. Lugli
e-mail: andrea.lugli8@unibo.it

M. Fiani
Dipartimento di Ingegneria Civile, University of Salerno, Via Giovanni Paolo II,
132, 84084 Fisciano, SA, Italy
e-mail: m.fiani@unisa.it

M. Scaioni (ed.), *Modern Technologies for Landslide Monitoring and Prediction*,
Springer Natural Hazards, DOI 10.1007/978-3-662-45931-7_5

differencing the DTM's. This analysis has been separately carried out in the different parts on which the landslide bodies had been subdivided.

Keywords Landslides · Monitoring · Terrestrial laser scanning · Digital terrain model · Geo-referencing

1 Introduction

Landslides are natural hazards of significant impact worldwide (Vervaeck 2013). In relation to slope lithology and morphology, several kinds of landslides exist, see Cruden (1991), Cruden and Varnes (1996), and Hungr et al. (2014).

A variety of physical processes could be responsible for triggering landslides, such as rainfalls, changes in slope hydro-geology, changes in slope geometry, earthquakes, rock/soil weathering, and anthropogenic activities, see www.safeland-fp7.eu. Among other geophysical and geotechnical techniques, the *geodetic methods* may give a fundamental contribution to the knowledge of both the surface topography and the kinematics of landslides, providing data useful for geologists, geomorphologists, and geotechnical engineers for interpreting possible triggering mechanisms of slope and rock failures. In the analysis of landslides (prediction, monitoring, and alerting for early warning, monitoring for change detection, etc.), a variety of geodetic techniques can be used, ranging from the most known and consolidated to the innovative ones. Geodetic techniques may provide three-dimensional information of the ground surface with various levels of accuracy and spatial resolution, enabling new ways of investigating the landslide phenomena. The choice of the method to be used depends on the morphology of the slope, on its extent, and on the typology of the landslide that has to be characterized or monitored. Any techniques present some critical issues and strengths as far as landslide monitoring applications are concerned, depending on the specific characteristics of the considered site.

Traditional surveying instruments like theodolites, global navigation satellite system (GNSS) receivers, and optical leveling allow the knowledge of the position (in 3D for the theodolites and GNSS, in 1D for optical leveling) of a few single control points with very high precision. Consequently, these instruments are suitable for precise monitoring of a landslide through a limited number of well-defined control points, but do not yield a complete description of specific areas or the evaluation of mass movements and mobilized volumes of soil or rock.

Notwithstanding GNSS receivers and theodolites can also be used for surveying a large number of non-signalized points on the ground, the availability of *terrestrial laser scanners* offers an unprecedented opportunity to quickly measure the terrain topography at high resolution and precision. Monitoring the surface deformation of wide landslides could be also accomplished with interferometric synthetic aperture radar (InSAR) processing from satellite or ground-based data, but the displacement direction has to be along the line of sight (LOS) of the sensor, see Metternicht et al. (2005) and ESA (2009). In order to detect also the other spatial components of the displacement vector, it is necessary to know their spatial orientation or to consider SAR image series acquired from ascending and descending orbits.

Apart from classical differential InSAR techniques, slow-moving landslides could be monitored also with *multi-temporal InSAR techniques* (MIT—Wasowski and Bovenga 2014) that allow a great precision in movement detection when several scenes are processed (Colesanti and Wasowski 2006; Fastellini et al. 2011). Among MIT, persistent scatterers InSAR (PSI) considers the evolution over time of a number of scatterers that retain the 'coherence' (in a few words, 'coherence' could be defined as the degree of similarity between corresponding pixels in multitemporal scenes) along a stack of repeated SAR images, while small baseline subset (SBAS) could be considered as an evolution of classical differential interferometry, allowing an unwrapping in time of the interferometric phase. In general, all InSAR techniques are not suitable for vegetated landslides, since vegetation does not allow sufficient coherence.

A promising technique is *ground-based InSAR* (GBSAR) (Monserrat et al. 2014), which could allow high precisions in the detection of movements along the LOS (Teza et al. 2007, 2008; Luzi et al. 2006; Lingua et al. 2008; Bozzano et al. 2011). GBSAR, however, presents some limiting factors:

- Such as satellite InSAR; it is not suitable for vegetated slopes;
- For distances greater than a couple of kilometers, the atmospheric effects become not negligible at all;
- The equipment is still very expensive; and
- Despite the millimeter accuracy in displacement detection, spatial resolution is low (about 50 cm, depending on the distance from the sensor).

Airborne laser scanning (ALS) provides precise numerical models of wide areas, with the capability of filtering vegetation, thanks to the multiecho beam (in particular if the surveying instrument is mounted on an helicopter, which allows flying at low altitudes, e.g., a few hundreds of meters), see McKean and Roering (2004), Barbarella and Gordini (2006).

High-resolution stereoscopic satellite images (Kliparchuk and Collins 2011) allow a good level of accuracy in the production of digital terrain models (DTM's) of the whole slope and therefore to monitor wide areas within a revisiting time of a few days. This technique may have further developments with the declassification of imagery at a spatial resolution of 25 cm, decided by the US Government on June 2014.

In this chapter, we describe the approach we used for TLS surveying and data processing in a 'real-world' case study, i.e., a large landslide that presents many criticalities in the design of a monitoring project over time. Our aim is to present a methodology that is 'sustainable' so as to provide a number of numerical and graphical products useful to allow both the geomorphological interpretation of active landslide phenomena and to monitor the surface shape changes occurred over time.

1.1 Terrestrial Laser Scanning

Among the geodetic surveying techniques, terrestrial laser scanning (TLS) has become quite popular for landslide monitoring owing to the easy use and the ability to collect a huge amount of data at high spatial resolution in short time.

Data coming from TLS can be processed for producing accurate and precise DTM's of surface topography and then to provide a precise and detailed description of geomorphology, see Slob and Hack (2004), Carter et al. (2001), Slatton et al. (2007), and Heritage and Large (2009). A TLS application to monitor a slow-moving landslide, removing returns from vegetation using a decision tree filter, is reported in Pirotti et al. (2013).

TLS seems indeed to be very suitable for landslide deformation measurements, especially for terrains with difficult access and steep slopes, since repeated surveys enable rapid 3D data acquisition of the topographic surface.

A detailed overview of LiDAR techniques applied to landslides, including TLS, is reported in Jaboyedoff et al. (2012). Main applications to landslides range from mapping and characterization (Derron and Jaboyedoff 2010; Guzzetti et al. 2012; Abellán et al. 2014) to monitoring (Abellán et al. 2009, 2010; Prokop and Panholzer 2009).

To achieve reliable and accurate results in landslide characterization and monitoring using TLS, it is necessary to follow a careful survey methodology and processing pipeline. In the case of rockslides, an interesting and exhaustive review is given by Abellán et al. (2014), which provides some principles for the quality assessment of TLS outputs, suggesting which operational steps have to be preferably accomplished.

The design of the survey is determined by the nature of the landslide to be monitored: its size, typology, speed, and geomorphology. The presence of obstacles may result in occlusions which often require the acquisition of multiple scans in order to complete the survey of the whole area, see Giussani and Scaioni (2004) and Soudarissanane et al. (2008).

Moreover, it is necessary to link the landslide DTM to some benchmarking areas which are stable over time; otherwise, the result of monitoring cannot be correctly interpreted. Indeed, in the case of reference points affected by displacements, the measurement of deformations will be influenced by the apparent movement due to geo-referencing errors.

Surveying of multiple point clouds, partially overlapped, requires *coregistration* as a first step in data processing. To carry out this specific task, a number of procedures have been implemented in scientific and commercial software packages, such as a rigid-body transformation of one point cloud into the reference system of another, by means of natural or artificial targets, or the well-known and effective iterative closest point (ICP) algorithm developed by Besl and McKay (1992), and with a different scheme for the research of corresponding points, improved by Chen and Medioni (1992). This algorithm aims at minimizing the 3D distance between partially overlapped surfaces in a set of 3D point clouds in order to improve the best image alignment with respect to the other 3D point clouds. Many variants of ICP have been proposed for improving the computational efficiency and to make automatic the pre-alignment stage. A detailed examination of the various approaches is given in Grün and Akca (2005), Monserrat and Crosetto (2008), and Pomerleau et al. (2013).

A limiting factor that greatly reduces the precision of point clouds derived from TLS surveying and specific processing algorithms is the presence of vegetation. While trees represent only an obstacle to the visibility of the terrain from the point of view of the laser scanning station, bushes and low vegetation can hide the actual shape of the soil.

Vegetation filtering step consists in the classification of data into terrain and off-terrain points, and due to the huge amount of 3D points acquired by the scanner, there is the need to automate the process. A detailed overview of several filtering algorithms that have been developed for this task and a representative selection of some methods are reported in Briese (2010).

The up-to-date TLS instruments allow the acquisition of *multiple echoes*. For example, Riegl VZ (www.riegl.com) series instruments are *full-waveform* systems. They can provide, if properly equipped with software tools, a full-waveform analysis very effective for filtering out the overgrown vegetation layer (Mallet and Bretar 2009; Elseberg et al. 2011; Guarnieri et al. 2012).

An interesting approach, in the attempt to effectively filter out vegetation, is represented by the analysis of a vegetation index computed by means of the combination of spectral bands gathered by the integrated RGB camera and a modified low-cost near-infrared camera, see Alba et al. (2011).

Data editing requires a large amount of work, but it is essential to extract the bare soil data from the whole laser scanning data set. Without this operation, it is not possible to use such data for a quantitative precise analysis of terrain displacements.

The next step of data processing chain involves the conversion of scattered point clouds into a regularized digital elevation model (DEM) that can be arranged in a grid data structure. Elevation values at the grid nodes have to be generated by appropriate interpolation methods, see Kraus and Pfeifer (2001), Vosselman and Sithole (2004), Kraus et al. (2006), and Pfeifer and Mandlburger (2009).

The accuracy of the DEM and its ability to faithfully represent the surface depends on the terrain morphology, the sampling density, the adopted interpolation algorithm, and on the grid spacing, see Aguilar et al. (2005) and Barbarella et al. (2013).

A DEM is appropriate for describing the surface of the landslide and to study its characteristics. Such model can be also exploited for deriving standard contour lines, but also to generate other important by-products like slope, aspect, and curvature maps.

If the goal of the survey is to measure ground deformations over time, two or more DEM's have to be compared (Fiani and Siani 2005; Abellán et al. 2009). Here, a number of different approaches can be used, see also Scaioni et al. (2014). For example, one can directly compare the DEM's obtained over time by simply fixing a number of points belonging to particular objects visible in the two different point clouds, see Ujike and Takagi (2004). Then, the estimated transformation parameters can be used to transform all the points of a cloud in the reference system of the other, making possible a comparison between them, see Hesse and Stramm (2004).

A more general approach allowing a full three-dimensional analysis useful for point cloud comparison is based on the algorithm called 3D least squares surface matching (LS3D) proposed in Grün and Akca (2005).

2 Case Study and Surveying Operations

The landslide we are considering is taking place in Pisciotta (Campania Region, Italy), on the left side of the final portion of a stream. This slope instability already resulted in extensive damage to both an important state road and two sections of the railway mainline connecting North and South of Italy.

Not long time ago, the railway has been interrupted by landslide dam-break just upstream: mud and soil brought by the landslide dam-break into the river have caused the blockage of both tunnels and the interruption of rail traffic on the mainline. In addition, a long stretch of the road appeared to be completely disrupted.

From a geological point of view, in accordance with the widely accepted classification of landslides already mentioned in Sect. 1, the active Pisciotta landslide is defined as a landslide compound system with a deep seated main body affected by long-term roto-translation, slow displacements. On surface, secondary earth slide flows with intermediate sliding rupture surface and shallow, rapid earth flows occurred.

It is highly possible that the geomorphological evolution of the landslide will cause the lowering of the steep slope above the road and the swelling of the lower part up to the bed of the stream. Such deformations drastically would disrupt the morphology of the basal part of the landslide where the railway bridges and the tunnel gates are located.

From the point of view of the geodetic surveying, the setup of the instruments is easier because the stream separates the two flanks of the valley: The right one is considered stable, while the left one is unstable. The upper parts of the two flanks are far away several 100 m from each other, while the lower parts are less than 100 m (Fig. 1).

Unfortunately, from the top of the stable part, the toe of the landslide cannot be seen. On the other hand, the slope toe is of great interest because in this area the slid debris accumulates. The TLS scans should therefore be acquired from multiple stations placed both on the upper part of the stable slope, from which the visibility toward the opposite site is very wide, and at the bottom, in order to make a very detailed survey of the lower part. The vegetation limits the choice of the placement of the instrument standpoints because of the presence of a few isolated trees. The visibility of the bare soil of the upper part of the unstable slope is, however, prevented because of the presence of shrubs.

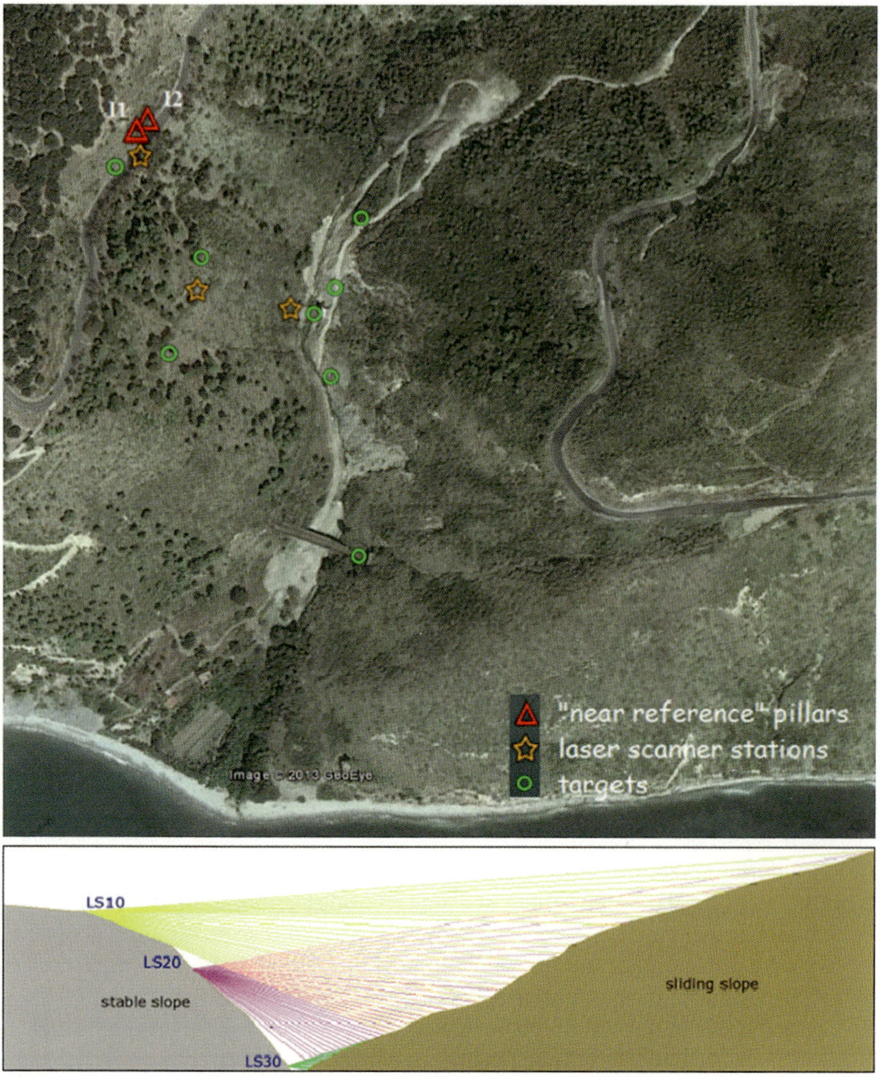

Fig. 1 *Panel above* planimetric positions of the 'near-reference' pillars (*red triangles*), the TLS stations (*orange stars*), and the targets (*green circles*) taken in a typical survey campaign. *Panel below* altimetric scheme

For some years, the part of the landslide surrounding the road was monitored (by a specialized company on behalf of the authority for national road management) by using fixed sensors, like weather stations, piezometers, inclinometers, strain gauges, and by geodetic surveying. This was carried out by materializing a couple of pillars (denoted in the following as I1 and I2) located on the top of the stable slope and measuring with theodolites the position of several control points materialized on the

sliding slope. In the period 2005–2009, two survey campaigns per month were carried out. The results of the periodic surveys showed that the amount of planimetric displacements in a period of about 4 years resulted in about 7 m, while the altimetric displacements were in the range between 2.5 and 3.5 m. The average horizontal daily speed of the landslide was approx. 0.5 cm/day, with peaks up to 2 cm/day. In elevation, it resulted in approx. 0.2 cm/day with peaks up to 1.5 cm/day. The displacements were referred to the couple of pillars (I1, I2) placed on the opposite flank of the valley. With this kind of survey, it was possible to determine quite accurately the displacements of any single points and to obtain information on the kinematics of the process, but it was impossible to compute the correct amount of the moving material or to precisely define the area affected by the landslide.

The use of TLS would enable the acquisition of large amounts of data and therefore potentially the knowledge of small details of the landslide. The goal of the survey campaigns that were run from 2010 to 2012 was to numerically describe the terrain morphology of the landsliding slope and its variation over time. For this purpose, we used the TLS surveying methodology in order to obtain a reliable 3D model of the landslide topographic surface (DTM) for each epoch and to derive the mass movements of the slope landslide, see Barbarella and Fiani (2013).

The rugged morphology of the surveyed area has been a strong constraint during the choice of the sites where the stations could be established. In all survey campaigns, we recorded laser scan measurements acquired from several TLS stations, located on the stable slope at different altitudes, in order to measure both the upper and the lower parts of the landslide body. Here, the shorter distances involved permitted the use of a medium-range laser scanner instrument, characterized by a higher measurement precision and spatial resolution that resulted in a larger point density.

At the toe of the landslide, there is a railway bridge connecting the two tunnels, and the terrain above the tunnel gate can be measured with high accuracy; not far from this, toward the valley, there is a longer bridge and a tunnel.

When the purpose of a survey is to monitor ground deformations, it is essential to identify a set of points that are stable in time and set up a stable reference system to which the repeated measurements over time could be referred.

In the case of a TLS surveying, if within the point cloud there is a sufficient number of recognizable and stable details, these may set up the reference frame and may be used to carry out the coregistration of the point clouds collected in subsequent measurement campaigns. If this is not the case, one needs to set up a reference system that remains stable over time, where the points measured in repeated surveys may be referred to. Such fixed points should be placed outside the monitoring area, but close enough to make easy and accurate the connection between them and the points to be surveyed on the landslide body.

Since in the case study the only recognizable and stable object is the tunnel gate, unsuitable to provide a stable frame, we used the couple of pillars previously materialized on the stable slope to set up the so-called near-reference frame. We carried out a GNSS survey in order to connect such pillars with some targets placed

on both sides of the valley and used them in order to geo-reference the TLS scans. The distance between targets and pillars ranged from a few dozens up to a few hundreds of meters.

As we need to be sure that the points belonging to the reference frame are stable over time, it was also advisable that such points were themselves monitored with respect to some additional reference points located in areas not affected by the landslide. An efficient and not expensive solution was to use as 'remote reference' frame a number of GNSS permanent stations (PS's) belonging to the IGS global network or to its local (national) densification, whose coordinates are known in a geodetic international frame.

The two pillars of the 'near-reference' frame were themselves connected to a couple of PS's belonging to a network real-time kinematic (NRTK) for real-time surveying service managed by the administration of Campania region. Such stations were located at 26 km (Castellabate) and 38 km (Sapri) far away from the landslide site, respectively. Due to the great distance between the landslide area and the PS's, we made continuous daily GNSS observations. By using them, we were able either to link the 'remote reference' stations to the 'near-reference' pillars or to calculate the GNSS coordinates of the targets placed on the landslide. Both the TLS stations and the targets were surveyed by GNSS geodetic-grade receivers in static mode and referred to the 'near-reference' frame, materialized by the two steel pillars firmly placed on a concrete wall in the stable area.

In Fig. 1, you may see both the survey setup and the measurement scheme for a 'typical' surveying campaign. We used different types of targets, mostly spherical with known diameter, but also cylindrical or planar as suggested by the manufacturers of the adopted laser instruments.

We carried out the first TLS survey campaign on February 2010. About 4 months later (June 2010), the second measurement campaign was accomplished. In both cases, a long-range instrument Optech ILRIS 36D was employed. A third surveying campaign was carried out one year later (June 2011) by using two different instruments, a long-range Riegl VZ400, and a medium-range Leica Scanstation C10 laser scanners. The second instrument was stationed on the lower part of the landslide (near the tunnel). The last campaign dated back to June 2012. Both Optech and Leica instruments were adopted by following the same organization of the previous surveys.

3 Data Processing

Data coming from TLS required several processing steps. Many of them were quite crucial in order to output reliable results. Once these steps were accomplished, a number of point clouds that described the landslide surface at different epochs were available.

GNSS data processing aims at determining the position of both targets and TLS stations in the selected reference framework. Since all the surveys have been

registered into the same reference system, it will be possible to compare the DTM's derived from the point clouds obtained at different epochs.

3.1 GNSS Data Processing

The GNSS data adopted during the surveying operations concerned the Global Positioning System (GPS) only. We can separate the elaboration of GPS data in three processing steps.

The first step concerned the connection between the two PS's we used, 26 and 38 km far away from the landslide, respectively, taken as fixed points and the two pillars of the 'near-reference' frame. The downloading of the PS observations acquired during the survey campaign allowed us to calculate the long GPS baselines between both pillars and the PS's.

The next step was the baseline adjustment. In such a way, we were able to obtain the adjusted coordinates of the pillars in the chosen geodetic datum that was the European terrestrial reference frame (ETRF00). This reference frame was selected because of its high stability over time (intraplate speed of the order of a few millimeters per year), see Altamimi et al. (2012a, b).

The third step of the GPS data processing concerned the computation of the short baselines between both targets and TLS stations and pillars, which were assumed as fixed at this stage. The length of these baselines ranged from 200 to 800 m approximately.

It is very important to acquire redundant observations to perform a rigorous least squares adjustment. In our case, two blunders in the last survey were detected by using 'data snooping' Baarda's procedure, see Teunissen (2000). Table 1 shows the mean values (in centimeters) obtained for the standard deviations of the adjusted coordinates of the two pillars I1 and I2, and a summary of the values obtained for the standard deviations of the targets.

The level of accuracy we have achieved for both the surveys of 2011 and 2012 is lower than the one for the two 2010 surveys. This is probably due to the discontinuous signal acquisition by the network NRTK in 2011 (which suggested not to use the point I2), and the noise induced by vegetation growth. The standard deviations of the position of the targets, however, are on average smaller than a couple of centimeters in planimetry and about twice in altimetry.

3.2 TLS Data Processing

The result of the four surveying campaigns carried out on the unstable slope consisted in a huge amount of data. In any survey campaign, we acquired from three to four scans taken from different stations. The scans are partially overlapped to help the alignment and to guarantee the coverage of the whole landslide (more

Table 1 Standard errors on pillar and target coordinates as computed from the GPS data processing

(cm)		Pillars		Targets		
Epoch	Std. dev.	I1	I2	Mean	Std. dev.	|max|
February 2010	σ_N	0.4	0.3	0.2	0.1	0.4
	σ_E	0.4	0.3	0.2	0.1	0.4
	σ_h	0.8	0.7	0.6	0.3	0.9
June 2010	σ_N	0.3	0.2	0.2	0.2	0.6
	σ_E	0.3	0.2	0.2	0.2	0.5
	σ_h	0.7	0.6	0.6	0.5	1.3
June 2011	σ_N	0.4	–	2.1	3.1	10
	σ_E	0.4	–	1.4	1.2	3.4
	σ_h	0.9	–	3.1	2.5	7.3
June 2012	σ_N	0.8	0.9	1.7	1.7	1.6
	σ_E	0.7	0.5	1.5	1.4	1.4
	σ_h	2	1.2	4.9	4.3	4.3

than 10^6 m^2). In the last two campaigns, a scan from the lowest station was also acquired using a short-range instrument characterized by higher scanning rate.

By processing all these data sets, we were able to obtain a number of DEM's of the landslide area at different epochs. This allowed us to make a comparison between them in order to evaluate the landslide evolution.

Processing of TLS data followed multiple steps. Alignment of overlapping scans is a very crucial step to achieve a good metric quality of the output. One should take special care in the scan pre-alignment phase, especially in the choice of the overlapping area between adjacent point clouds and in the selection of the homologous points for coregistration. This task is mathematically operated by using a six-parameter rigid-body transformation. In a first phase, a pre-alignment was obtained from the manual selection of a few corresponding points. Then, a best fit alignment was carried out with ICP algorithm. We applied standard registration procedures based on point matching. The points in the scans that have been selected by the operator are used by the algorithm as starting points for the alignment. We used both natural points and targets as tie points.

Both the software packages PolyWorks (Innovmetric) and Cyclone (Leica GeoSystems) use the least squares iterative algorithm ICP to optimally align a set of 3D images by minimizing the 3D distances between the points of overlapping portions of the 'reference' point cloud and the 'data' point cloud. The strategy used by RiscanPro software package in order to minimize these errors is very similar, since the software has a plug-in function called 'multistation adjustment' (MSA) that tries to improve iteratively the registration of the scan positions. To compare the scan positions, both tie points, tie objects, and poly-data objects (reduced point clouds) are used.

A crucial step in TLS data processing consists in removing those points that have not been reflected from the ground but from vegetation or obstacles, as well as the outliers caused by ‘noise.’ To isolate the points that belong to the bare soil from others is important especially in deformation measurement applications, where the real ground movement must be detected.

In order to filter out the vegetation and to edit data, we used different software tools according on the adopted laser instrument. Almost all the software packages implement semi-automatic filtering procedures. They are often based on the same algorithms that have been designed for aerial laser scanning (ALS) applications, see Axelsson (2000) and Vosselman and Sithole (2004).

The editing tools that we used followed different approaches: Some of them are based on manual inspection, while others implement semi-automatic procedures. We used PolyWorks version 10.0 (Innovmetric) to edit both the February and June 2010 scans and Cyclone to edit the 2011 scans from Leica C10.

While PolyWorks uses a completely manual procedure, in Cyclone a semi-automatic approach can be applied for slicing the point cloud in a number of stripes of length specified by the operator. On any single stripe, thanks to the visualization in 2D, we can manually delete those points that do not belong to the bare soil. Such a procedure is very time-consuming, yet the result is quite accurate. The thinner the strip, the better the result.

In Fig. 2, we show the 2D views of a number of stripes having different thickness. Editing is still to be done. We edit the whole scan by using strips 5 m large.

For editing the 2012 scans acquired from Riegl Z400, we used RiscanPro version 1.2 (Riegl) editing tool. We applied two different spatial filters that are implemented in the software package: ‘octree’ and ‘iterative filtering.’

Our experience teaches us that great care must be put in the use of semi-automatic procedures because there is a risk of accidentally deleting some portions of soil. Only by comparing the surfaces derived from different epochs, some interpretation errors may be detected.

Once an overall cloud that represents the bare soil is available, it should be registered in the chosen absolute reference system by measuring the target coordinates. All software packages allow the geo-referencing of point clouds. The procedure consists in assigning a set of 3D coordinates to a number of points that can be recognized in the point cloud.

In particular, when there are no terrain details that can be considered stable and that are placed in a suitable position within the point cloud, a precise geo-referencing is a quite critical task. For monitoring purpose, a correct geo-referencing in a reference system that is stable over time allows us to make an effective comparison between scans in an absolute reference system.

In our case study, no natural features located on the stable slope or in the surrounding area were recognizable in the point cloud, to be used to set up a reference frame stable over time. Consequently, some artificial targets were adopted. Their position with respect to the chosen reference system is known since they have been measured on the ground.

Software tools generally allow the automatic recognition of the target shape (cylindrical, spherical, or flat) at least on the type of targets suggested by the

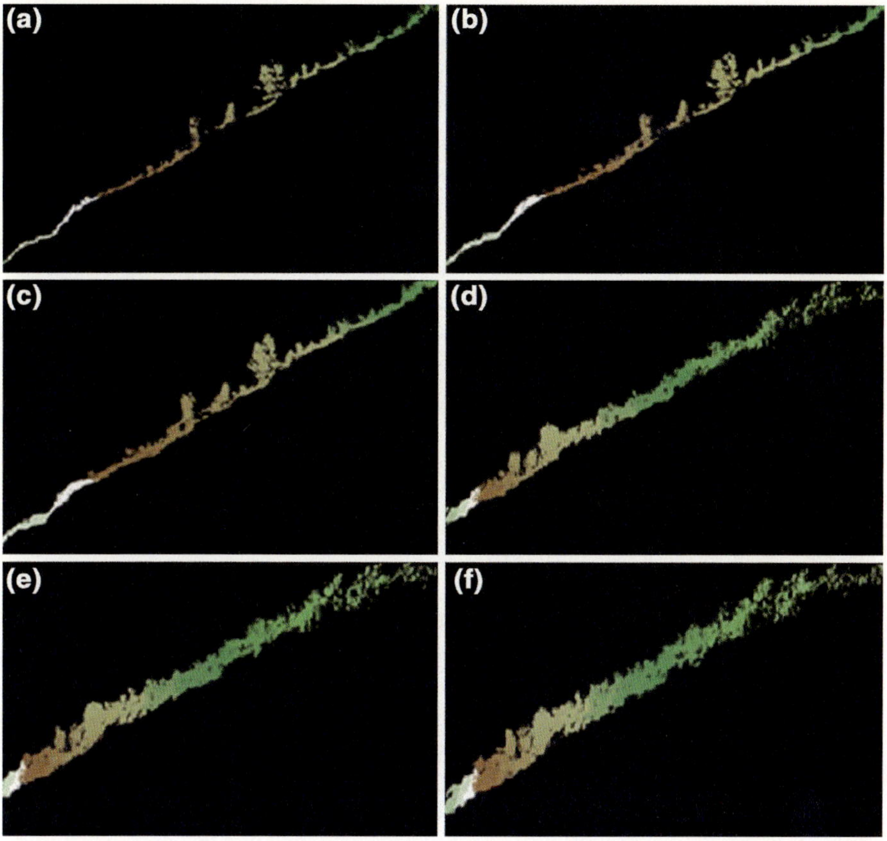

Fig. 2 View in 2D of some examples of *vertical* stripes having different thickness **a** 1 m; **b** 2 m; **c** 3 m; **d** 5 m; **e** 8 m; **f** 10 m

manufacturer of the associated TLS instrument. In order to accurately estimate the position of the center of the targets, we need to border the portion of points of the target to be used for the fitting calculus.

We made a seven-parameter (similarity) transformation in order to geo-reference all the scans. A six-parameter rigid-body transformation gives almost the same results. We note that usually the laser data processing software packages do not provide a rigorous analysis of the quality of the standardized residuals of the transformation. The process of geo-referencing has led to values of 3D residuals ranging from 5 cm to 6 cm.

We computed the geo-referencing of all the surveys in the ETRF00 Geodetic System. In order to separate more easily the planimetric from the altimetric components, we transformed the planimetric coordinates in the cartographic system UTM_ETRF00. For the altimetry, we continued to use the ellipsoidal heights. In the following, we assumed y = North; x = East; z = height.

There are many algorithms to generate a DEM of a surface starting from a point cloud. Most of them are designed to produce a grid DEM. In the landslide analysis, the choice of the algorithm is not trivial, depending on various factors as terrain morphology and presence of discontinuities. The assessment of the surface deformation occurred over time is strongly influenced by the DEM generation mode.

Once all the raw DTM's, i.e., the geo-referenced and filtered point clouds, have been built, we gain a number of numerical models of the surface of the landslide at different epochs (four in our test case) that can be compared between them. Since all the surveys have been framed in the same reference system, such a comparison allows us to estimate the landslide displacements.

3.3 DEM Processing

Starting from the geo-referenced point clouds, we were able to derive a number of DEM's which described the whole surface of the landslide body and allowed us to analyze the deformation occurred over time.

In order to calculate the height to be given to each node of the grid DEM, one of the many existing interpolation algorithms had to be selected. The choice of the interpolator greatly may influence the accuracy of the DEM. In landslide characterization and monitoring, the choice of both the interpolation algorithm and the grid step is not trivial, since their choice depends on various factors such as the point density and distribution pattern, the terrain morphology, and the presence of breaklines.

To study the ability of some different algorithms to reconstruct a surface adherent to the data, a preliminary test was accomplished in a test area. Such region featured a quite complex morphology but did not host any overgrown vegetation. The test point cloud consisted of about 4 million points. Both an artifact (the gate of a tunnel) and sharp discontinuities were in the test data set, see Fig. 14 in Sect. 5.

For evaluating the adherence of the DEM interpolated surface to the input point cloud, we considered the height difference between the laser points and the interpolated surface, i.e., the difference between the height z_k of the point and the interpolated height $z_{\text{int}}(x_k, y_k)$, in the same planimetric position. We applied a cross-validation method, extracting from the cloud a test sample of 1 % of points and then calculating the height differences between the points belonging to this sample (containing a number of points $N = 40{,}000$) and the interpolated surface generated by means of the remaining 99 % of the original data. Thereby, the points of the test sample did not contribute to the construction of the DEM.

For the modelling of the surface, we tested several well-known interpolation algorithms, which are widely used in surface modelling (Mongillo 2011) and implemented in commercial software packages as Surfer (Golden Software) and ArcMap (ESRI):

- Inverse distance to a power, second degree (Idw2);
- Kriging, linear variogram (Kri);
- Local Polynomial, first degree (LoPo1°) and second degree (LoPo2°);
- Natural Neighbor (NatNe); and
- Radial Basis Function, multiquadric, smoothing factor $c^2 = 0.0001$ (RBF0001), $c^2 = 0.0013$ (RBF–sug), $c^2 = 0.01$ (RBF01).

A few checkpoints gave very large residuals for all the algorithms tested. These points are mainly located in areas characterized by abrupt breaklines such as the tunnel wall, the steep slopes, and the like. Those residuals exceeding 1 m in absolute value were considered as outliers and discarded from the testing sample. Their percentage varies within the tested algorithms, ranging from the smallest values of 3.8 % for Idw2, 3.9 % for Kri and RBF0 to a maximum of 5.5 % for RBF-sug (the value of smoothing factor suggested by the adopted software package).

Once the outliers were removed, the frequency of the residuals at intervals of 5 cm was computed to assess the level of closeness between the points and the DEM. Figure 3 shows, for all the nine algorithms we tested, the curve of the frequency of absolute residuals exceeding a threshold. For both the algorithms IDW2 and LoPo2, the 5 % of the absolute residuals exceeded 14 cm, while for the algorithm RBF01, the 5 % of the absolute residuals exceeded 24 cm.

As a global measure of a DEM's accuracy, Yang and Hodler (2000) proposed the use of the *root mean square error* (RMSE) of the residuals, a parameter that can be easily statistically interpreted if the residuals follow a normal distribution. However, the samples of the residuals seem to be characterized by a different behavior, regardless of the type of interpolator used. When intervals of 5 cm have been considered and we computed the frequency of absolute residuals x for each interval, we obtained the histogram shown in Fig. 4, built for the results obtained

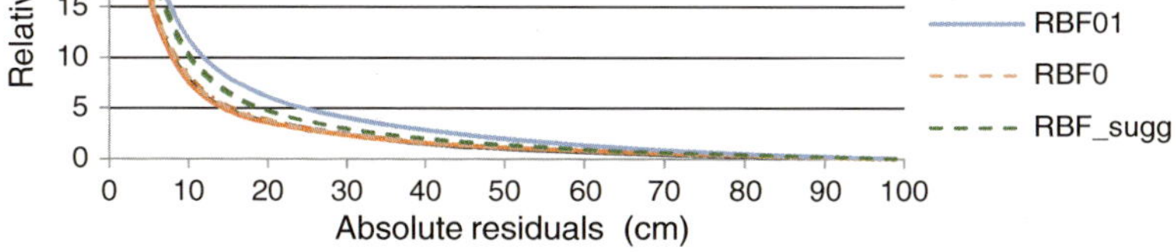

Fig. 3 Frequency of the absolute residuals that do not exceed a given threshold (after removing outliers, i.e., points with absolute residuals larger than 100 cm)

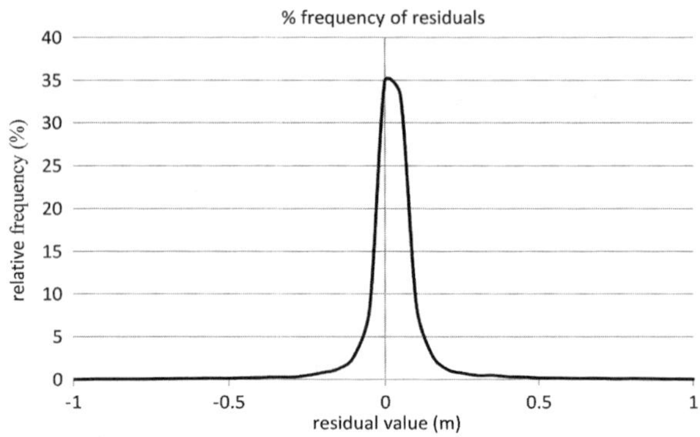

Fig. 4 Step histogram of the relative frequency of residuals on 5-cm interval

when using the Idw2 algorithm. The trends we obtained from the other algorithms were almost similar to the one in Fig. 4.

It was very hard to consider the residuals of the interpolations to be normally distributed, regardless of the algorithm tested.

To interpret the statistical behavior of the data sample, we normalize the residuals considering some other parameters, namely mean, RMSE, median, skewness ($g1$), kurtosis ($g2$), and median absolute deviation (mad).

The mathematical expressions of the statistical parameters adopted for normalization are the following ones:

$$m_x = \frac{1}{N}\sum_{i=1}^{N} x_i \quad \text{rmse} = \sqrt{\sum_{i=1}^{N} (x_i - m_x)^2/(N-1)} \tag{1}$$

$$\text{mad} = \text{median}(|x_i - \text{median}(x)|) \tag{2}$$

$$g_1 = \frac{1}{N}\sum_{i=1}^{N} (x_i - m_x)^3 / \left(\frac{1}{N}\sum_{i=1}^{N} (x_i - m_x)^2\right)^{3/2} \tag{3}$$

$$g_2 = \frac{1}{N}\sum_{i=1}^{N} (x_i - m_x)^4 / \left(\frac{1}{N}\sum_{i=1}^{N} (x_i - m_x)^2\right)^2 - 3 \tag{4}$$

The numerical values of the parameters computed on the test data when considering different interpolation algorithms are shown in Table 2.

The statistical parameters of the various algorithms look similar: The value of RMSE ranges from 13 to 15 cm, the mad ranges from 2.9 to 3.7 cm, and the mean value is always positive and less than 1 cm. If we compare the algorithms in terms of the ability to describe the terrain, the best results are given by Idw2, LoPo2°, and RBF0 ($c^2 = 0$). The relative differences between them are nevertheless quite small.

Table 2 Some statistical parameters for the tested interpolation algorithms

Algorithm	m_x (cm)	s_x (cm)	mad (cm)	g_1	g_2	s_x/mad
Idw2	0.3	13.1	2.9	0.4	17.8	4.5
Kri	0.5	13.5	2.9	0.5	17.8	4.7
LoPo1	0.5	13.3	2.9	1.0	18.3	4.6
LoPo2	0.5	13	2.8	0.8	18.8	4.6
NatNe	0.5	13.5	2.9	0.6	18.0	4.7
RBF0	0.5	14.7	3.3	0.6	22.2	4.5
RBF-sugg	0.3	15.2	3.7	0.2	13.3	4.1

Mean (m_x), root mean square error (s_x), median absolute deviation (mad), skewness (g_1), and kurtosis (g_2); in addition, the rightmost column reports the ratio of rmse and mad; results are given in cm except for g_1, g_2, s_x/mad, and s_x/mad that are dimensionless

As a result of our tests, we chose the Idw2 algorithm that gave us the best overall performance.

This interpolation method allowed us to introduce the breaklines that are needed to better characterize the trend of the ground especially in the case of artifacts, see Lichtenstein and Doytsher (2004). At a glance, we built the grid DEM's used for the subsequent analysis of movements by using the Idw2 algorithm. We drew the breaklines just in correspondence of the mouth of the tunnel, which is the unique artifact in the scene.

4 Results

In order to follow the evolution of the landslide morphology under investigation, it was devised to consider as reference the surface derived from the survey in February 2010. For this reason, as a convention, given two surveys acquired at different epochs, differences between the corresponding DEM's were always calculated by subtracting the latest to the oldest surfaces.

The landslide process was studied through three methods: (1) points *versus* mesh comparison, (2) DEM subtraction, and (3) cross-sectional analysis. Results are discussed in the following subsections.

4.1 Points Versus Mesh Comparison

In laser scanner software, surface generation is obtained as a *mesh* through a Delaunay triangulation (this means that it is not possible to choose which interpolator to use, like in a GIS environment).

The 'Compare' procedure, as implemented in PolyWorks, consists in measuring the distance of each point of a point cloud with respect to the mesh obtained from the other point cloud, chosen as reference. Here, the distances along the shortest path from the point to the mesh were considered. The sign of the distance depends on the relative position of the point cloud and the mesh: If a point is above the mesh, the distance will be positive (in this case, it means that some debris accumulated); otherwise, if a point is below the mesh, the sign will be negative (this means erosion). At the end of the process, input data points are represented in a chromatic scale whose entities correspond to the distance values.

Comparisons between the four point clouds have been carried out by considering as a reference surface the interpolated mesh in February 2010. The computation of distances between single points of a generic epoch and the mesh surface is suitable for a first look identification of stable/unstable areas, both on the whole slope (Fig. 5) and across small areas. February 2010 scan covers an area slightly wider than June 2012; this is the reason of the gray parts in the figure.

It has, however, to be clearly pointed out that the result of a 'Compare' procedure is not a volume variation calculation.

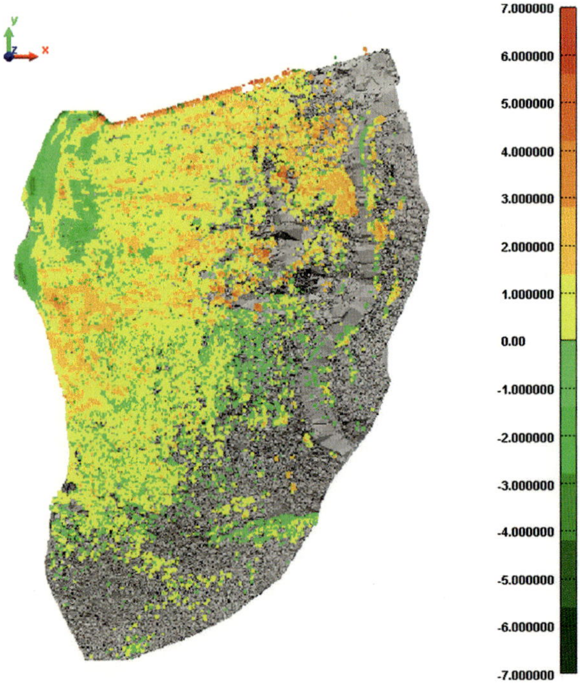

Fig. 5 Comparison based on the distance between points and mesh surface (see Sect. 4.1) of the point cloud gathered in June 2012 with respect to the reference mesh obtained from the point cloud in February 2010, which is represented in *gray*. Points with colors from *dark* to *light green* correspond to erosion areas, points from *yellow* to *brown* to accumulation areas

From an operational point of view, in PolyWorks, it is possible to perform calculation of volume variation only in more than one step:

- First, it is necessary to create an horizontal plane to be taken as a reference;
- Then, for each of the two mesh surfaces, the volume with respect to this reference plane is computed using the 'Surface-to-Plane Volume' tool; and
- Finally, volume difference between the two mesh is obtained indirectly by manually differencing the volume results previously read.

Moreover, working in vectorial domain, PolyWorks does not allow direct map algebra operations. For these reasons, it has been chosen to calculate volumes directly in the raster domain, using a GIS software.

4.2 DEM Subtraction

In the study of a landslide, the evaluation of terrain mobilized during a certain period is very important. In commercial software packages oriented to TLS data processing, this result is often achieved by differencing the volumes comprised between the interpolated surfaces and a horizontal reference plane. In order to get a quantitative estimation of mobilized volumes in terms of loss and gain, here it was chosen to perform this computation in a GIS environment, i.e., ArcMap 10.1 (ESRI).

The grid DEM's derived from both scans were used for differencing any couples of epochs. It was considered a cell spacing of 50 cm, devised accordingly to the mean span of the points in the fours surveys. Consequently, all the derived products were obtained from the DEM's interpolated with this cell spacing. Differencing the heights between cells of corresponding DEM's provided the information about height variations over time.

Results obtained by differencing the last survey (June 2012) and the first survey (February 2010) are shown in Fig. 6. In Fig. 7, we show the DEM differences in the interim periods. This output should be equivalent to the one shown in Fig. 5, even though different palettes (due to different software packages) and adopted measures (shortest distance between any points and the mesh, difference along the vertical of the elevation values in correspondence of grid nodes) were used.

Since the most interesting information is that about volume variations, it is advisable to perform such calculations by considering homogenous areas with respect to the relative position (below/above) of the surfaces. To distinguish areas in erosion from those of accumulation, the resulting grid has been classified into three classes, see Fig. 8:

- From −0.1 to 0.1 m: areas which could be considered as characterized by the balance of loss and gain (for sake of simplicity in the following, these points will be defined just as 'stable');

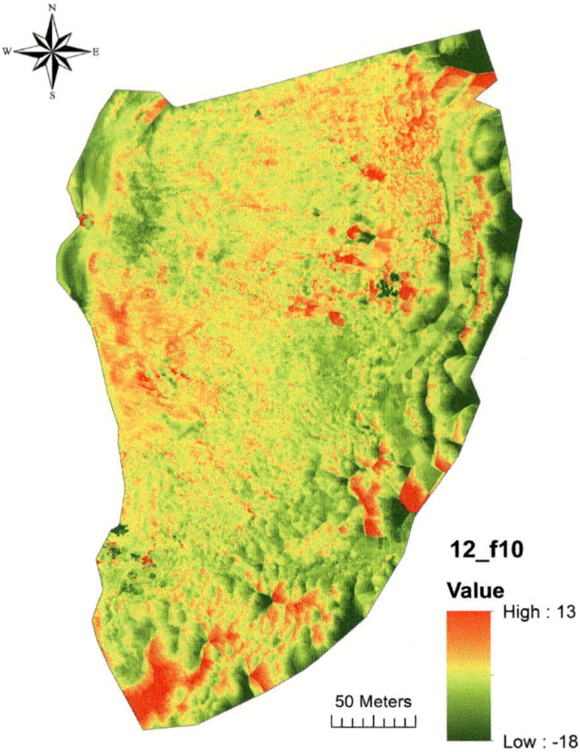

Fig. 6 Difference map obtained by DEM subtraction method (see Sect. 4.2) in February 2010 and June 2012; results are displayed in meters

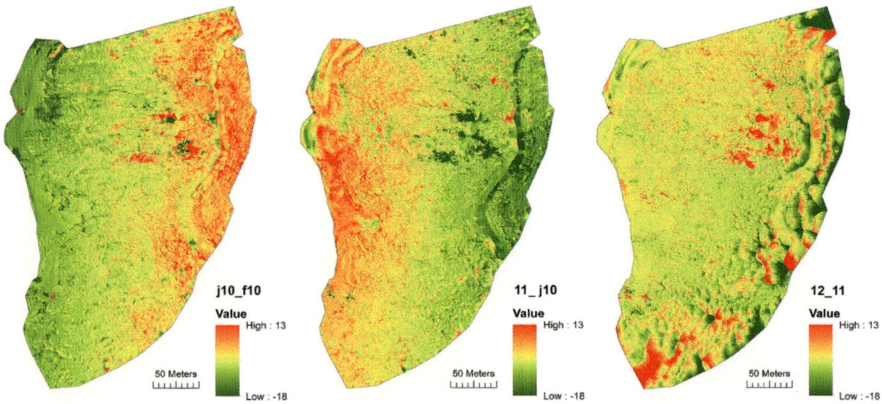

Fig. 7 Subtraction of DEM's in the periods February 2010–June 2010, June 2010–June 2011, and June 2011–June 2012 (from *left* to *right*, respectively)

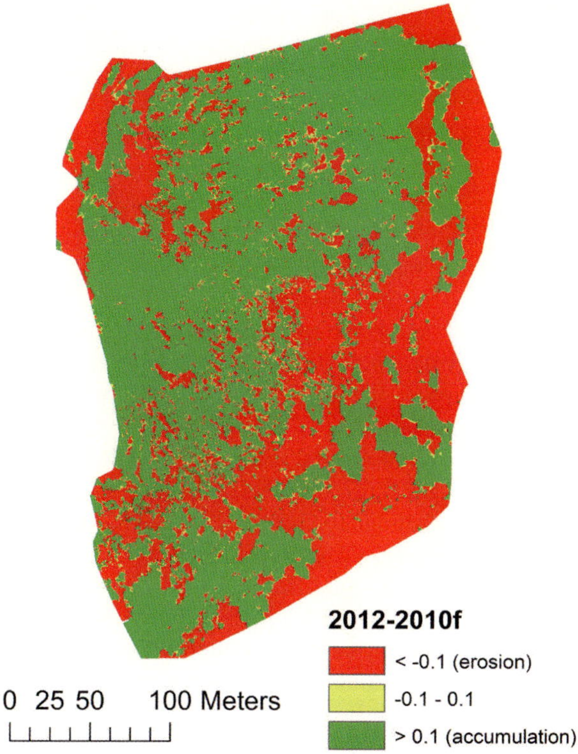

Fig. 8 Automatic segmentation into three classes of the results from grid DEM subtraction obtained in the case of February 2010 and June 2012

- Equal or greater than 0.1 m: areas in accumulation from February 2010 to June 2012; and
- Equal or lower than −0.1 m, areas in erosion.

The volumes mobilized over the years have been calculated on the entire part of the landslide common to the four surveys, resulting from the intersection of all the boundaries. In Fig. 8, in green are represented cells corresponding to accumulation, in red those corresponding to erosion, and in yellow those that could be considered as stable.

This method allowed determining that from February 2010 to July 2012, the mobilized volume resulted as 58,820 m^3 of total accumulation and 48,851 m^3 of total erosion, mainly due to excavation. The total volume corresponding to points considered as stable (from −0.1 to 0.1 m of difference) was about 235 m^3.

The term 'erosion' here should not be interpreted only as landslide displacement or as weathering, but also as an effect of artificial removal of landslide debris which could dam the stream and the nearby gravel road.

Even if this classification into three classes represents a sort of rough segmentation, this approach is reductive because information related to the global volume does not describe the relative displacements between single portions of the landslide.

Considering the landslide as a whole body does not allow an analysis of the differential kinematics between the several bodies of the landslide.

4.2.1 Landslide's Bodies' Identification

To make a classification of the different homogeneous parts of the landslide, the advice of geomorphologists played a fundamental role. Experts of the landslide area, indeed, were able to identify and draw the different polygons to be analyzed on the basis of the visual inspection of map output from laser scanning surveys.

More in detail, the geomorphologists were provided with contour maps with small intervals, up to 0.25 m (Fig. 9 left, shown with a contour interval of 2 m in order to improve the readability of the figure), slope and aspects maps (Fig. 9 center and Fig. 9 right).

The geomorphological interpretation allows us to recognize the landslide bodies shown in Fig. 10 (Guida 2013, personal communication).

The morphological analysis showed that the slope under study (orographic left bank) is affected by landslides that differ in type, age, and stage of activity.

On the main body of the landslide, indeed, it is possible to recognize landslides of second and third generation. The continuous progress of the landslide gradually shrinks the bed of the creek, either along the horizontal or the vertical directions.

It is possible to identify six polygons that represent the following:

- A landslide of second generation indicated by the abbreviation '1_3' in Fig. 10;
- Two landslides of third generation indicated by the abbreviations '1_1_2' and '1_1_3'; and
- Three landslides of fourth generation classified as '1_1_0_1', '1_1_2_1', and '1_1_3_1.'

The main body of the landslide, which would have been indicated by the abbreviation '1', is not represented in Fig. 10, as well as the third-generation

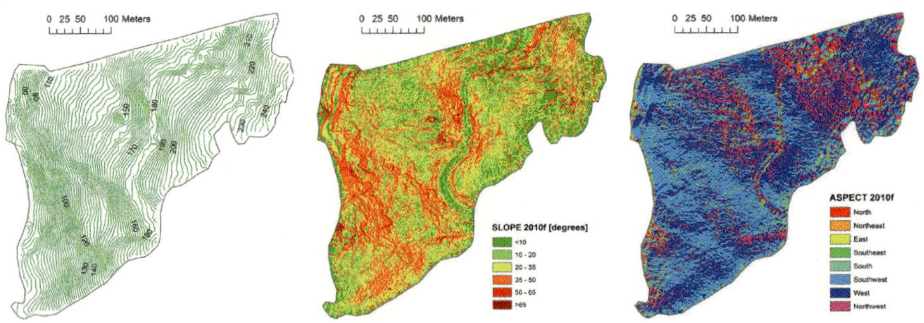

Fig. 9 From *left* to *right* contour line map with vertical interdistance of 2 m; slope map; aspect map. All maps were derived from the TLS survey in February 2010

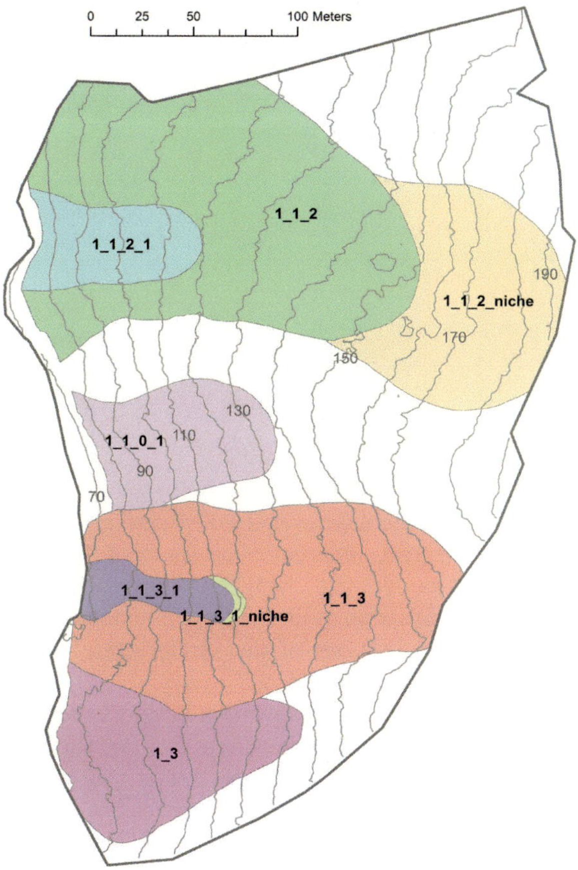

Fig. 10 Landslide bodies identified considering the survey in February 2010, overlapped to the corresponding contour lines (drawn with an interval of 10 m). The shapes of landslide bodies have been clipped with respect to the boundary common to the four surveys (represented in *dark gray*)

landslide body classified as ‘1_1_0’, because their extension exceeds the area detected by some of the acquired TLS surveys.

With regard to these different landslide bodies identified throughout the landslide, volume calculations have been performed considering only two classes, i.e., erosion and accumulation. More in detail, for each landslide body have been calculated both its aggregated volume and both volumes corresponding to erosion and accumulation zones. Results of this analysis are reported in Table 3.

In general, there is a good correspondence between volume losses and gains among the several landslide bodies. For instance, the niche ‘1_1_2’ shows a volume loss/gain balance which could be related to the volume gained by the areas below.

Moreover, human interventions that took place in 2011 in order to stabilize the toe of the landslide do not allow a clear interpretation of its behavior: Area ‘1_1_2_1’ shows a total loss of about four hundreds of cubic meters in spite of a noticeable loss of the above landslide bodies.

Table 3 Mobilized volumes from February 2010 to July 2012, expressed in cubic meters; positive values mean accumulation, negative ones mean erosion

Landslide body/mobilized volumes (m^3)	10 J_10 F	11_10 J	12_11 J
1_1_0_1	5,720	−1,350	440
1_1_2	22,192	−19,680	6,660
1_1_2_1	1,392	−2,080	530
1_1_2_niche	26,000	−24,220	−2,210
1_1_3_1	1,580	−650	−230
1_1_3_1_niche	270	−90	−60
1_3	6,890	−5,900	4,880
1_1_3	24,850	−19,240	−3,860

4.3 Cross-Sectional Analysis

Along with contour level and DEM-derived maps shown in Fig. 9, cross sections are tools of great interest for the study of the morphology of a landslide and for monitoring topographic changes over time.

A quantitative analysis of the variations in terrain surface is also possible to compare cross sections of the surfaces resulting from the interpolation of point clouds at different epochs.

A critical aspect of this method is represented by the choice of the orientation of the cross-sectional profile along the landslide, which basically defines which components of the displacement vector can be observed. Indeed, a wrong orientation of the profile could lead to unnecessary or misleading interpretation. For this reason, in comparing cross sections taken at different positions along the slope, it should be considered that the landslide orientation could have changed between each other.

Here, it has been devised to take profiles along the axes of landslide bodies, measuring the differences along the vertical direction between the four profiles. From February 2010 to July 2012, the variation in height (thickness) measured in some points has nearly reached 3.5 m, as can be seen in Fig. 11.

Of course, an exhaustive description of the whole landslide would require a lot of vertical cross sections, representing therefore a time-consuming operation. Indeed, while the computation of each profile is an automatic procedure given the DEM, the selection of the location and the orientation of any profiles require careful analyses. For these reasons, the extraction of cross sections is limited to a few narrow areas where a deeper investigation is needed, rather than to be worked out on the entire slope.

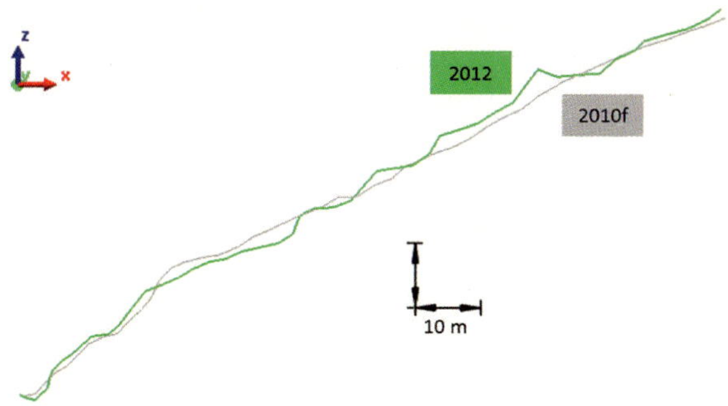

Fig. 11 Profiles of the slope interested by the landslide, from the top to hillside, taken into consideration epochs in June 2012 and February 2010. The section corresponds to the axis of landslide body '1_1_2_1', which is in the nearby of the right side of the railway tunnel

5 Discussion

Among TLS applications, landslide monitoring is one of the more complex because it is not limited to the description of a surface but requires also quantitative comparisons, essential to determine the kinematics of the phenomenon and for planning subsequent works on the ground. As a matter of fact, the morphology of a slope does not obviously correspond to a well-defined geometric pattern, recognizable in its positions at different epochs.

The final data to be processed—the shape of a surface obtained from a point cloud—are derived from a series of field operations and a processing pipeline that necessarily involves approximations and interpolations of the original data and that may affect the comparison of the surfaces.

As pointed out by several authors (e.g., Abellán et al. 2014), it is therefore necessary to perform TLS surveys by carefully following a set of procedures for data processing and calculation.

Some critical issues concern, for instance, the problem of geo-referencing the survey into a frame stable over time and how to perform the comparison between surveys acquired at later ages.

The significance of the identified displacements depends indeed on the reliability of the framing of the different surveys in the same reference frame stable over time. In the case of the survey of stations and targets performed with a satellite GNSS method, it is advisable to consider a reference frame solid with the tectonic plate of the investigated area, even if the locations of the reference stations are usually far away from the 'near' reference, and therefore, the position of these points has already a not negligible uncertainty which propagates to rise up the standard deviation of the surveyed points. As a matter of fact, what is apparently loose in precision is largely gained in reliability of the absolute positioning.

It has, however, to be taken into account that the precision of point cloud position obtained from a survey depends also on the coregistration and geo-referencing steps. An assessment of the overall geo-referencing precision of the several point clouds can be operated by exploiting the presence of artifacts that are stable over time and thus allow the evaluation of the survey repeatability. In the application reported here, these are the tunnel and the railway bridge, which are continuously monitored on site with an interferometer because of their relevant function. The interferometric measurement did not show any movements in the observation period.

Thanks to the very high point density, allowed by the proximity of the slope with respect to one of the stations, it was possible to identify in the four surveys several corresponding details on the vertical wall (oriented north–south) of the tunnel.

The average differences of the locations of these 'homologous' checkpoints with respect to the first survey on February 2010 are reported in Table 4.

Since the total shift observed is always less than 10 cm, it does not seem appropriate to apply any translation on the basis of this local result. To assess the repeatability of the height component, it is more convenient to consider the height differences in the points belonging to the horizontal plane of the bridge.

Figure 12 shows the difference between June 2012's and February 2010's DEM's. Contour levels corresponding to DEM subtraction are shown in Fig. 13.

Table 4 Shifts of checkpoints with respect to the first survey in February 2010

Shift versus 2010 f (cm)	2010 J	2011 J	2012 J
Est	−1.9	−0.5	6.3
Nord	−1.8	9.0	−3.5
Height	−0.3	−0.4	−3.8
Total	2.6	9.0	8.1

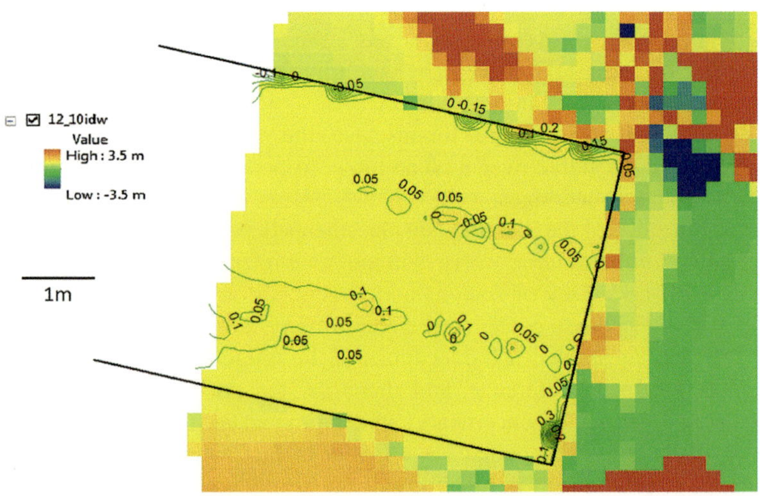

Fig. 12 Contours extracted from the 2012-to-February 2010 difference grid, considering the tunnel main geometries as breaklines (in *black*), with a contouring interval of 5 cm

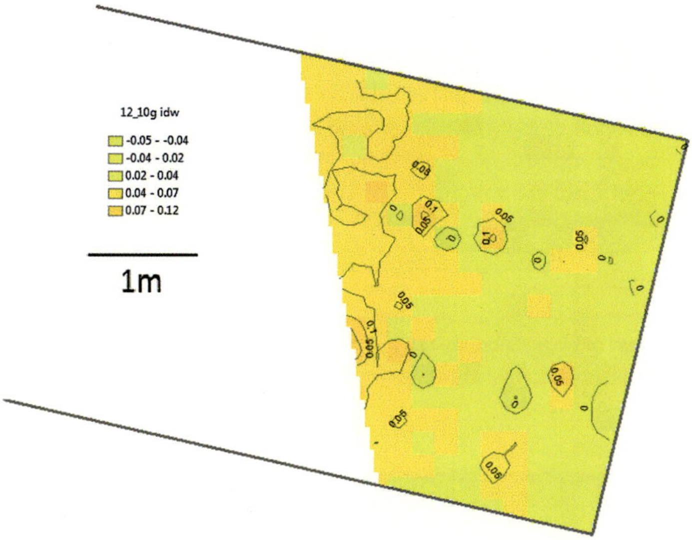

Fig. 13 Contours extracted from the 2012-to-June 2010 difference grid; here have been interpolated only points comprised inside railway tracks. For this reason, the palette is different from the one in Fig. 12

Here, the interpolation was carried out by considering the main geometries of the railway tunnel as breaklines (in black). In the figure, it has been chosen a contour interval of 5 cm and to represent in yellow difference values around zero.

Contour lines corresponding to height differences of 5 and 10 cm lie almost exclusively in correspondence of the tracks and are therefore to be considered probably linked to the different acquisition geometry of the crossties (e.g., support elements) of the tracks in the two surveys. The area around the tunnel has not been considered in calculations, in order not to affect the results with spurious differences deriving from the different acquisition geometry of any scans.

As already stated, there are various ways for comparing surveys acquired at different epochs; the choice of which method to use depends on the particular context and on the object being monitored. Moreover, details are often better interpreted considering cross sections than comparisons or grid differences. Consider for instance the retaining wall in the area nearby the tunnel, see Fig. 14.

The comparison between scans acquired in February 2010 and in June 2012 does not allow clear results, as shown in Fig. 15, which shows a detail of the retaining wall. This is probably because the surface of the reference mesh is continuous and more or less the blocks are positioned along the steepness of the slope. Only a few anomalies are pointed out by some patches clearly different from the surrounding areas, with irregular shapes not corresponding with the geometry of an ashlar. Also the map obtained from DEM subtraction, shown in Fig. 16, does not allow a better interpretation.

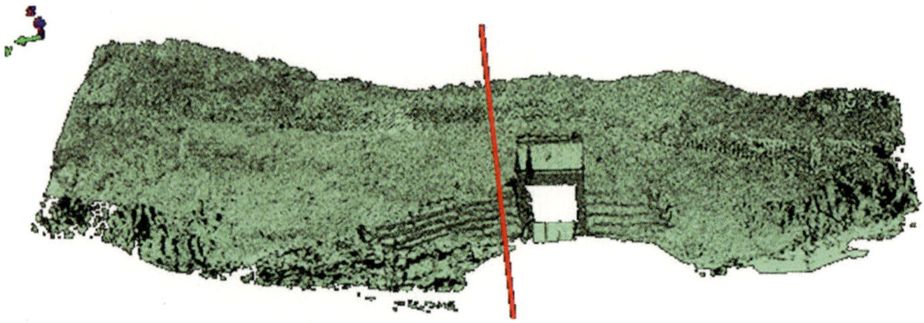

Fig. 14 Test area; mesh of February 2010 scan, in red the profile line taken perpendicular to the retaining wall

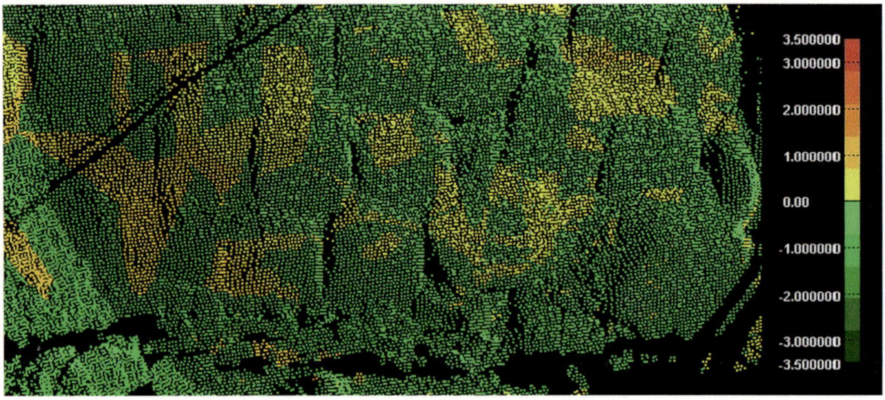

Fig. 15 Comparison of June 2012 point cloud with respect to February 2010 reference mesh, close-up over the ashlars nearby the railway tunnel

A clear evidence of human intervention in body '1_1_2_1' area can be found by comparing profiles taken over the four scans: The blocks of the retaining wall on the right side of the railway tunnel were removed and repositioned, passing from three to five rows. In Fig. 17, the cross sections taken along the blocks of the retaining wall close to the railway tunnel are shown (the position of the section line is indicated in red in Fig. 14). In better detail, an analysis of the cross sections points out that the intervention at the retaining wall occurred in the time elapsed between the June 2010 scan and the June 2011.

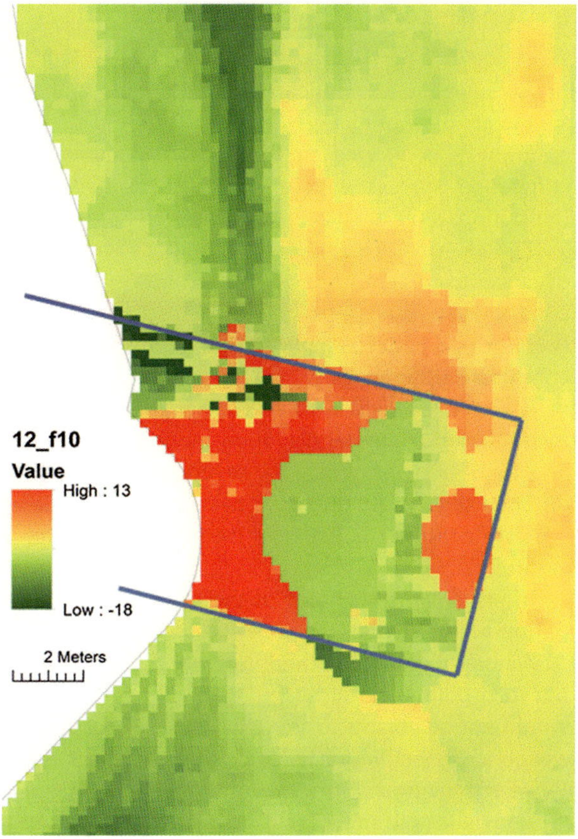

Fig. 16 Map obtained from DEM subtraction in the area of the railway tunnel portal

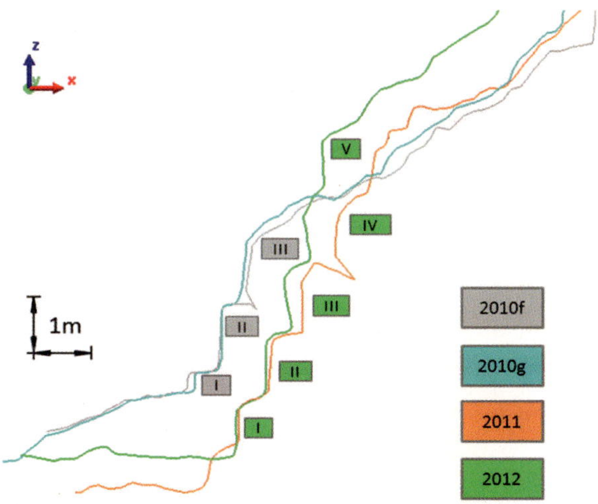

Fig. 17 *Cross sections* of the four scans; Latin numbers indicate the rows of ashlars of the retaining wall in February 2010 and in June 2012

6 Conclusions

In this chapter, an application of multitemporal surveys based on TLS techniques for the analysis of changes and deformations of a landslide has been presented. The case study is located in Pisciotta (Campania, Italy), where an active landslide compound system with a deep seated main body affected by long-term roto-translation, slow displacements, as well as secondary earth slide flows on surface.

The use of GNSS permanent stations as a frame stable over time, despite their distance from landslide area, has, however, allowed a reliable geo-referencing; controls on a stable artifact allow to assess that the repeatability level of positioning is comprised between 5 and 10 cm, corresponding to the threshold of significance of coordinate differences.

It is also advisable to make a preliminary analysis about how to perform the interpolation of digital elevation models (DEM's), because, given a certain density and distribution of points, the several available algorithms may return models with a different degree of adherence to the input data. It is symptomatic, for example, the great influence of the value of the smoothing factor when using radial basis function algorithm.

In principle, the choice of the method for surface comparison strongly depends on the nature of the phenomenon to be observed.

Among the three methods adopted for the monitoring of this landslide (distances between points and a mesh, DEM subtraction, vertical cross-sectional analysis), it would not be correct to state that there is one more appropriate than others because each one is suitable for appreciating a different aspect of the observed phenomenon. DEM subtraction allows volume calculations, also over wide areas, while cross sections allow to detect displacements occurred in a restricted area along a certain profile. The evaluation of mobilized volumes is not a trivial task and requires the expertise of geomorphologists in order to define homogeneous areas from geomorphological point of view.

References

Abellán, A., Jaboyedoff, M., Oppikofer, T., & Vilaplana, J. M. (2009). Detection of millimetric deformation using a terrestrial laser scanner: Experiment and application to a rockfall event. *Natural Hazards Earth System Science, 9*, 365–472.

Abellán, A., Vilaplana, J. M., Calvet, J., & Blanchard, J. (2010). Detection and spatial prediction of rock falls by means of terrestrial laser scanning modelling. *Geomorphology, 119*, 162–171.

Abellán, A., Oppikofer, T., Jaboyedoff, M., Rosser, N. J., Lim, M., & Lato, M. J. (2014). Terrestrial laser scanning of rock slope instabilities. *Earth Surface Processes and Landforms, 39*, 80–97.

Aguilar, F. J., Agüera, F., Aguilar, M. A., & Carvaj, F. (2005). Effects of terrain morphology, sampling density, and interpolation methods on grid DEM accuracy. *Photogrammetric Engineering & Remote Sensing, 71*, 805–816.

Alba, M. I., Barazzetti, L., & Scaioni, M. (2011). Filtering vegetation from terrestrial point clouds with low-cost near infrared cameras. *Italian Journal of Remote Sensing, 43*(2), 55–75.

Altamimi, Z. (2012a). ITRS, ETRS89, their relationship and Realization. In *Proceedings of the EUREF Symposium*, Paris Saint-Mandé, France.

Altamimi, Z., Métivier, L. & Collilieux, X. (2012b). ITRF2008 plate motion model. *Journal of Geophysical Research, 117*, paper no. B07402. doi:10.1029/2011JB008930.

Axelsson, P. (2000). DEM generation from laser scanner data using adaptive TIN models. *International Archives of Photogrammetry and Remote Sensing, 33*(B4/1), 110–117.

Barbarella, M., & Fiani, M. (2013). Monitoring of large landslides by terrestrial laser scanning techniques: Field data collection and processing. *European Journal of Remote Sensing, 46*, 126–151.

Barbarella, M., Fiani, M., & Lugli, A. (2013). Landslide monitoring using multitemporal terrestrial laser scanning for ground displacement analysis. *Geomatics, Natural Hazards and Risk*. doi:10.1080/19475705.2013.863808.

Barbarella, M., & Gordini, C. (2006). Kinematic GPS survey as validation of LIDAR strips accuracy. *Annals of Geophysics, 49*(1), 21–33.

Besl, P. J., & McKay, N. N. (1992). A method for registration of 3D shapes. *IEEE Transactions on Pattern Analysis and Machine Intelligence, 14*(2), 239–256.

Bozzano, F., Cipriani, I., Mazzanti, P., & Prestininzi, A. (2011). Displacement patterns of a landslide affected by human activities: Insights from ground-based InSAR monitoring. *Natural Hazards, 59*, 1377–1396.

Briese, C. (2010). Extraction of digital terrain models. In G. Vosselman & H. G. Maas (Eds.), *Airborne and terrestrial laser scanning* (pp. 135–167). Dunbeath, UK: Whittles Publishing.

Carter, W., Shrestha, R., Tuell, D., Bloomquist, D., & Sartori, M. (2001). Airborne laser swath mapping shines new light on earth's topography. *EOS Transactions of American Geophysical Union, 82*, 549–555.

Chen, Y., & Medioni, G. (1992). Object modelling by registration of multiple range images. *Image and Vision Computing, 10*(3), 145–155.

Colesanti, C., & Wasowski, J. (2006). Satellite SAR interferometry for wide-area slope hazard detection and site-specific monitoring of slow landslides. *Engineering Geology, 88*(3–4), 173–199.

Cruden, D. M. (1991). A very simple definition for a landslide. *IAEG Bulletin, 43*, 27–29.

Cruden, D. M. & Varnes, D. J. (1996). Landslides types and processes. In A. K. Turner & R. L. Schuster (Eds.), *Landslides: Investigation and Mitigation, Transportation Research Board Special Report No. 247* (pp. 36–75). Washington, USA: National Academy Press.

Derron, M. H., & Jaboyedoff, M. (2010). Preface to the special issue: LIDAR and DEM techniques for landslides monitoring and characterization. *Natural Hazards Earth System Science, 10*, 1877–1879.

Elseberg, J., Borrmann, D. & Nuchter, A. (2011). Full wave analysis in 3D laser scans for vegetation detection in urban environments. In *Proceedings of the 23rd International Symposium on 'Information, Communication and Automation Technologies (ICAT)'*.

ESA. (2009). European space agency GMES. The Terrafirma ATLAS 2009. Available at: www.terrafirma.eu.com (Last Access on 28th August, 2014).

Fastellini, G., Radicioni, F., & Stoppini, A. (2011). The Assisi landslide monitoring: A multi-year activity based on geomatic techniques. *Applied Geomatics, 3*, 91–100.

Fiani, M. & Siani, N. (2005). Comparison of terrestrial laser scanners in production of DEMs for Cetara tower. In *Proceedings of the CIPA Conference*. Torino, Italy, 26 September–1 October 2005.

Giussani, A., & Scaioni, M. (2004). Application of TLS to support landslides study: Survey planning, operational issues and data processing. *International Archives of the Photogrammetry, Remote Sensing and Spatial Information Sciences, 36*(8/W2), 318–323.

Grün, A., & Akca, D. (2005). Least squares 3D surface and curve matching. *ISPRS Journal of Photogrammetry and Remote Sensing, 59*(3), 151–174.

Guarnieri, A., Pirotti, F., & Vettore, A. (2012). Comparison of discrete return and waveform terrestrial laser scanning for dense vegetation filtering. *International Archives of the Photogrammetry, Remote Sensing and Spatial Information Sciences, 39*(B7), 511–516.

Guzzetti, F., Mondini, A. C., Cardinali, M., Fiorucci, F., Santangelo, M., & Chang, K. T. (2012). Landslide inventory maps: New tools for an old problem. *Earth Science Reviews, 112*, 42–66.

Heritage, G. L., & Large, A. R. G. (2009). *Laser scanning for the environmental sciences* (p. 302). Chichester, UK: Wiley.

Hesse, C. & Stramm, H. (2004). Deformation measurements with laser scanners—Possibilities and challenges. In *Proceedings of the International Symposium on 'Modern Technologies, Education and Professional practice in Geodesy and Related fields'*, Sofia, Bulgaria.

Hungr, O., Leroueil, S., & Picarelli, L. (2014). The Varnes classification of landslide types, an update. *Landslides, 11*, 167–194.

Jaboyedoff, M., Oppikofer, T., Abellán, A., Derron, M. H., Loye, A., Metzger, R., & Pedrazzini, A. (2012). Use of LIDAR in landslide investigations: A review. *Natural Hazards, 61*, 1–24.

Kliparchuk, K., & Collins, D. (2011). Evaluation of stereoscopic geoeye-1 satellite imagery to assess landscape and stand level characteristics. *International Archives of The Photogrammetry, Remote Sensing and Spatial Information Sciences, 38*(1), 6.

Kraus, K., & Pfeifer, N. (2001). Advanced DTM generation from LIDAR data. *International Archives of the Photogrammetry, Remote Sensing and Spatial Information Sciences, 34*(3/W4), 23–30.

Kraus, K., Karel, W., Briese, C., & Mandlburger, G. (2006). Local accuracy measures for digital terrain models. *The Photogrammetric Record, 21*, 342–354.

Lichtenstein, A. & Doytsher, Y. (2004). Geospatial aspects of merging DTM with breaklines. In *Proceedings of FIG Working Week 2004 on 'NSDI and Data Distribution'* , Athens, Greece, May 22–27, 2004.

Lingua, A., Piatti, D., & Rinaudo, F. (2008). Remote monitoring of a landslide using an integration of GB-InSAR and LiDAR techniques. *International Archives of The Photogrammetry, Remote Sensing and Spatial Information Sciences, 37*(1), 361–366.

Luzi, G., Pieraccini, M., Mecatti, D., Noferini, L., Macaluso, G., Galgaro, A., & Atzeni, C. (2006). Advances in ground based microwave interferometry for landslides survey: A case study. *International Journal of Remote Sensing, 27*(12), 2331–2350.

Mallet, C., & Bretar, F. (2009). Full-waveform topographic lidar: State-of-the-art. *ISPRS Journal of Photogrammetry and Remote Sensing, 64*(1), 1–16.

McKean, J., & Roering, J. (2004). Objective landslide detection and surface morphology mapping using high-resolution airborne laser altimetry. *Geomorphology, 57*, 331–351.

Metternicht, G., Hurni, L., & Gogu, R. (2005). Remote sensing of landslides: An analysis of the potential contribution to geo-spatial for hazard assessment in mountainous environments. *Remote Sensing of Environment, 98*, 284–303.

Mongillo, M. (2011). Choosing basis functions and shape parameters for radial basis function methods. Available at: www.siam.org/students/siuro/vol4/S01084.pdf (Last Access on August 28th, 2014).

Monserrat, O., & Crosetto, M. (2008). Deformation measurement using terrestrial laser scanning data and least squares 3D surface matching. *ISPRS Journal of Photogrammetry and Remote Sensing, 63*, 142–154.

Monserrat, O., Crosetto, M., & Luzi, G. (2014). A review of ground-based SAR interferometry for deformation measurement. *ISPRS Journal of Photogrammetry and Remote Sensing, 93*(1), 40–48.

Pfeifer, N. & Mandlburger, G. (2009). LiDAR data filtering and DTM generation. In J. Shan, C. K. Toth (Eds.), *Topographic Laser Scanning and Ranging. Principles and Processing* (pp. 307–334). Boca Raton, FL, USA: Taylor & Francis Group.

Pirotti, F., Guarnieri, A., & Vettore, A. (2013). Vegetation filtering of waveform terrestrial laser scanner data for DTM production. *Applied Geomatics, 5*(4), 311–322.

Pomerleau, F., Colas, F., Siegwart, R., & Magnenat, S. (2013). Comparing ICP variants on real-world data sets. *Autonomous Robots, 34*(3), 133–148.

Prokop, A., & Panholzer, H. (2009). Assessing the capability of terrestrial laser scanning for monitoring slow moving landslides. *Natural Hazards and Earth System Science, 9*, 1921–1928.

Scaioni, M., Feng, T., Lu, P., Qiao, G., Tong, X., & Li, R., et al. (2014). Close-range photogrammetric techniques for deformation measurement: Applications to landslides. In M. Scaioni (Ed.), *Modern technologies for landslide investigation and prediction* (pp. 13–41). Berlin Heidelberg: Springer.

Slatton, K. C., Carter, W. E., Shrestha, R. L. & Dietrich, W. (2007). Airborne laser swath mapping: Achieving the resolution and accuracy required for geosurficial research. *Geophysical Research Letters, 34*(23), paper no. L23S10, 5 p.

Slob, S., & Hack, R. (2004). 3D terrestrial laser scanning as a new field measurement and monitoring technique. *Engineering geology for infrastructure planning in Europe: A European perspective* (Vol. 104, pp. 179–189). Lectures Notes in Earth Sciences Berlin Heidelberg: Springer.

Soudarissanane, S., Lindenbergh, R., & Gorte, B. (2008). Reducing the error in terrestrial laser scanning by optimizing the measurement set-up. *International Archives of The Photogrammetry, Remote Sensing and Spatial Information Sciences, 37*(5), 615–620.

Teunissen, P. J. G. (2000). *Testing theory. An introduction*. Series on Mathematical Geodesy and Positioning. The Netherlands: Delft University Press.

Teza, G., Galgaro, A., Zaltron, N., & Genevois, R. (2007). Terrestrial laser scanner to detect landslide displacement fields: A new approach. *International Journal of Remote Sensing, 28*, 3425–3446.

Teza, G., and others. (2008). Ground based monitoring of high risk landslides through joint use of laser scanner and interferometric radar. *International Journal of Remote Sensing 29*:4735–4756.

Ujike, K. & Takagi, M. (2004). Measurement of landslide displacement by object extraction with ground based portable laser scanner. In *Proceedings of the 25th Asian Conference on Remote Sensing*, Chiang Mai, Thailand.

Vervaeck, A. (2013). *Worldwide landslide report*. In Earthquake-Report.com, Available at: http://earthquake-report.com/2013/06/01/worldwide-landslide-report/ (Last Access on August 28th, 2014).

Vosselman, G., & Sithole, G. (2004). Experimental comparison of filter algorithms for bare-earth extraction from airborne laser scanning point clouds. *ISPRS Journal of Photogrammetry and Remote Sensing, 59*(1–2), 85–101.

Wasowski, J., & Bovenga, F. (2014). Investigating landslides and unstable slopes with satellite multi temporal interferometry: Current issues and future perspectives. *Engineering Geology, 174*, 103–138.

Yang, X., & Hodler, T. (2000). Visual and statistical comparisons of surface modelling techniques for point-based environmental data. *Cartography Geographic Information Science, 27*(2), 165–175.

Micro-scale Landslide Displacements Detection Using Bayesian Methods Applied to GNSS Data

Francesco Pirotti, Alberto Guarnieri, Andrea Masiero, Carlo Gregoretti, Massimo Degetto and Antonio Vettore

Abstract In this chapter, we evaluate the movement of 6 points near a landslide body, which were surveyed with GNSS receivers over time. We apply Bayesian inference to identify the areas on the ground with statistically significant vertical (downwards) shifts. Traditional statistical methods work well only when point displacements between different survey epochs are sufficiently large compared to the standard deviations of related coordinates. In such cases, coordinate differences of some points can be marked as potential displacements. The Bayesian analysis can help to improve discrimination when height differences, computed with respect to the first measurement epoch, are at the same order of magnitude as the uncertainties of the measures. After the application of the classical statistical test, one network point, close to the upper part of the landslide area, seemed to be more unstable than the remainder. In order to remove or validate the hypothesis of instability, the Bayesian statistical inference was applied, and all three of the upper group of points show significant shift, depending on the data *prior* parameters. This application shows that the Bayesian approach can be considered as an integration to

F. Pirotti (✉) · A. Guarnieri · A. Masiero · A. Vettore
CIRGEO-Interdepartment Research Center of Geomatics, University of Padova, Padova, Italy
e-mail: francesco.pirotti@unipd.it

A. Guarnieri
e-mail: alberto.guarnieri@unipd.it

A. Masiero
e-mail: andrea.masiero@unipd.it

A. Vettore
e-mail: antonio.vettore@unipd.it

C. Gregoretti · M. Degetto
TESAF-Department of Land, Environment, Agriculture and Forestry, University of Padova, Padova, Italy
e-mail: carlo.gregoretti@unipd.it

M. Degetto
e-mail: massimo.degetto@unipd.it

M. Scaioni (ed.), *Modern Technologies for Landslide Monitoring and Prediction*, Springer Natural Hazards, DOI 10.1007/978-3-662-45931-7_6

classical statistical significance testing (e.g. z-test), reliably showing significance in vertical directional (i.e., downwards) coordinate shifts, thus supporting detection of movements having lower magnitude.

Keywords Bayesian methods · Deformation measurement · GNSS control networks · Landslides · Monitoring

1 Introduction

1.1 Geodetic Techniques for Change Detection of the Earth Surface

There are various techniques to monitor movements of the earth surface. Movement, by definition, has a spatial and a temporal component, two aspects which have to be measured accurately to successfully carry out tasks related to monitoring these in these two domains. Besides traditional geodetic and geotechnical surveying methods (GNSS, robotic theodolites, boreholes, inclinometers, etc.) adopted for investigation and monitoring of landslides, the use of modern remote sensing techniques for the study of these phenomena has exponentially grown in last years. The spatial component can be measured directly using classical topographic techniques or indirectly by estimating movement using remote sensing and related geomatic techniques, i.e. photogrammetry (see Scaioni et al. (2014b)). The advantages of the latter methods are evident especially when one considers the low degree of accessibility of landslide areas and the high degree of risk for personnel that carry out the direct measurements. The disadvantages are related to the resolution, accuracy and capability of the sensors; for example, if vegetation is present over the area, photogrammetry alone will not provide ground information, whereas *light detection and ranging* (LiDAR) allows a certain penetration of the canopy and thus returns the information on the ground plane (Pirotti et al. 2013). Indeed, the possibility of acquiring highly detailed and accurate *digital terrain models* (DTM's) offered by *ground-based interferometric synthetic aperture radar* (GBSAR) (see Monserrat et al. (2014)) and LiDAR techniques (Dowman 2004), has opened new way of applications for the study of landslide phenomena (Lingua et al. 2007). In this field, GBSAR-based systems are mainly used for the detection and quantification of small displacements over large areas (Crosetto et al. 2014; Monserrat et al. 2014; Refice et al. 2000; Ye et al. 2004). Specific case studies regarding *synthetic aperture radar and permanent scatterers* have been tested by Farina et al. (2006) and Frangioni et al. (2014). GNSS techniques with 'low-cost' receivers have been investigated and have attracted interest, for understandable economic reasons (Cina and Piras 2014). Forlani et al. (2013) use GNSS for camera positions in the 'Photo-GNSS' technique which applies dense-matching algorithms to retrieve surface models of the terrain (see Previtali et al. (2014) and Remondino et al. (2014)). A *spatial sensor network* (SSN) was tested in Scaioni et al. (2014a) including photogrammetry and contact geotechnical sensors on

a scaled-down model of a landslide simulation platform. *Terrestrial laser scanners* are also becoming popular for multi-temporal change detection of landslides since Bitelli et al. (2004) especially considering that latest models have increased range and decreased weight of the sensor and the aforementioned capability to filter out vegetation (Pirotti et al. 2013). Other references are given in Barbarella et al. (2013, 2014), Jaboyedoff et al. (2012).

All the methods mentioned are valid, carry intrinsic pros and cons and hence have to be carefully applied depending on the characteristics of the phenomenon that is being studied, i.e. the landslide. A state-of-the-art network of GNSS receivers remains a most robust approach to detect micro-scale displacements.

1.2 GNSS for Deformation Monitoring of Earth Surface

Geodetic techniques are widely used for monitoring the deformation of the earth surface at different spatial and temporal scales. The term global navigation satellite system (GNSS) is used to define positioning systems based on a constellation of satellites, which emit carrier signals used for defining time and position of the receiving station. Various systems are either operational or about to be so. The most recognized, due to longer operation time, is the US GPS constellation followed by Russia's GLONASS, Europe's Galileo and the Chinese Beidou (see Hofmann-Wellenhof et al. (2007)). Precise positioning is fundamental for monitoring dynamics of elements on the earth and to support other geo-spatial technologies, e.g. remote sensing and geographic information systems (GIS) which together act in synergy for the assessment of natural hazards and risk (see Manfré et al. (2012)). GNSS estimates of position and derivatives (i.e. velocity and acceleration) are becoming more reliable, and advanced data analysis techniques are helpful for the recognition of features in the GNSS time series, e.g. nonlinear behaviours, discontinuities in the signal and in its derivatives. Detection of signal discontinuities between two or more GNSS multi-temporal surveys, or in a whole time series, can be accomplished through the use of advanced analysis techniques such as *wavelets*, the *Bayesian* and the *variational methods* (see Borghi et al. (2012)). Discontinuities which are expected to be very small and compatible with the signal noise motivate the use of advanced data analysis techniques to investigate significant modifications of point positions (see Betti et al. (1999)). The ability to detect GNSS points whose movement can be considered significantly different from noise and other factors not related to the phenomenon of interest is an important step in the field of geomatics applied to natural hazards and risk. Investigations in this sense have been carried out also by Wang and Soler (2012) where a GNSS dense network was monitored to detect a creeping landslide in a 2-year period; authors also discuss influence of rainfall events which degrade performance of the receivers. The Bayesian approach in structuring a significance test (Koch 2007) on the displacement of point position is a robust and promising approach. Important work on this topic has also been carried out by Betti et al. (2001, 2011).

In this case study, we will present a proof-of-concept using a network of GNSS receivers to determine significance of the movement recorded on the vertical axis. All measures by sensors have a budget of residual errors, and each error function can be estimated and used in a model to calculate significance. This is typical in statistical analysis, and literature provides us with several methods. In the following chapter, we present and discuss the Bayesian approach used to analyse a time series of vertical displacement measures carried out in a GNSS network with ten vertices. The objective is to assess significance of each measure testing different models that represent the characteristics of the phenomenon under investigation, thus allowing better discrimination with respect to classical statistical inference.

2 Study Area

The area which was tested is called Rovina di Cancìa, located in the Dolomites region (Northeast Italy—Fig. 1). It includes a basin whose main channel origins at the feet of Salvella Fork (2,500 m a.s.l.) in the municipality of Borca di Cadore (Fig. 2), 15 km far from Cortina d'Ampezzo and ends in a retaining basin systems (low deposition area) at 1,005 m a.s.l. The channel intersects a flat area at lower altitude (1,344 m a.s.l.) which was specifically built to divert and slow down debris flow. Phenomena of landslide and of debris flows are present at this site. The latter are triggered just downstream a cyclopean boulder (Fig. 3) where run-off entrains large quantities of solid material and debris flow can form. For more detailed discussion on the site (see Gregoretti and Dalla Fontana (2008)).

3 Materials and Methods

3.1 The GNSS Control Network

A GNSS control network (CN) was created around the landslide area in order to determine small surface movements. The CN consists of ten points, four of which (1–4) are considered stable and are positioned at a distance from the landslide (Fig. 1), whereas the remaining 6 points (5–10) are positioned near the landslide body. The fixed points were defined using the Italian geodetic network from the Istituto Geografico Militare (IGM 2014) which has monumented several control points in the region. With classical topography (total station), we defined the fixed points where the GNSS receivers were then positioned.

The location of the set of the control points to monitor was detected using orthophotos and a LiDAR-derived digital terrain model (DTM) from an aerial survey carried out in 2003. The DTM was used to take into account the terrain morphology, logistics and safety (slopes steepness) issues, thus the overall

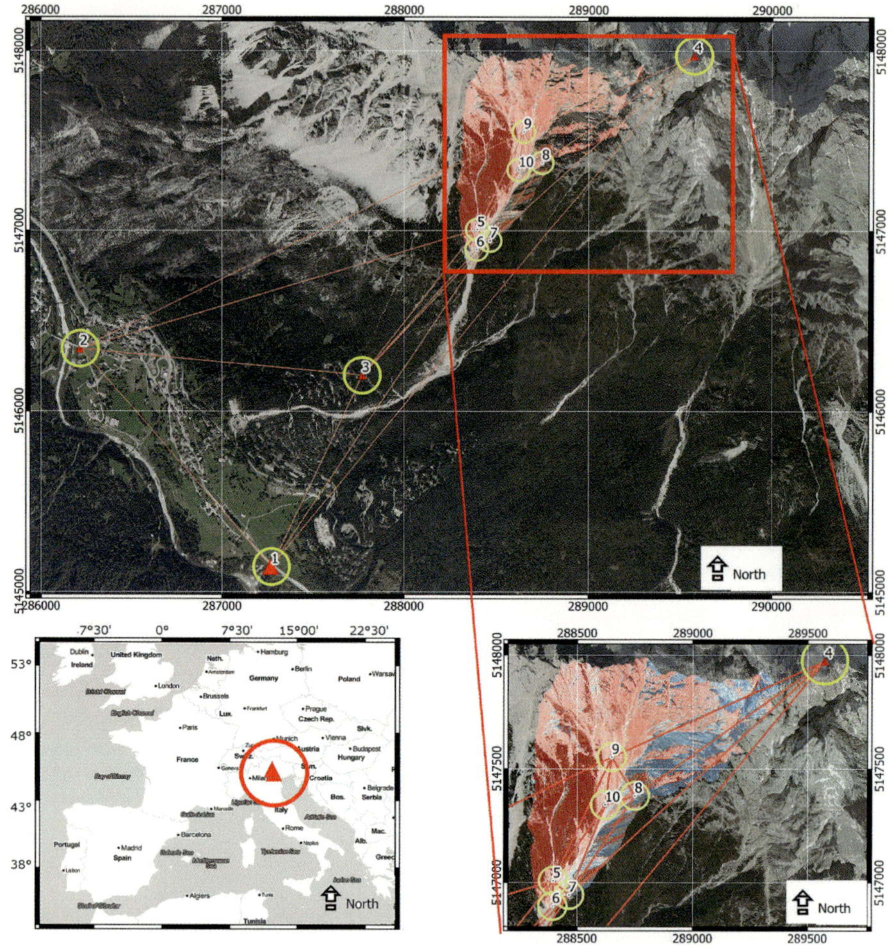

Fig. 1 The GNSS control network

accessibility. In order to determine the best set of candidate locations, several sky-plots were determined. Potential obstructions (e.g. due to vegetation, rocks and overall morphology) were derived from the LiDAR-derived DSM as well as from visual interpretation of the ortho-images. Five benchmarks were then selected and properly monumented in the field by cementing steel survey nails into the ground or into the rocks. This solution was adopted in order to easily recover the marks in subsequent surveys and to prevent possible displacements due to damage by local fauna or cattle. The length of all potential baselines of the resulting geodetic network ranges between 300 m and 5 km. Four survey campaigns have been performed between 2011 and 2013, with a 6-month time interval: in May, after the snow-melting period and in early October, before the winter season. Occupation

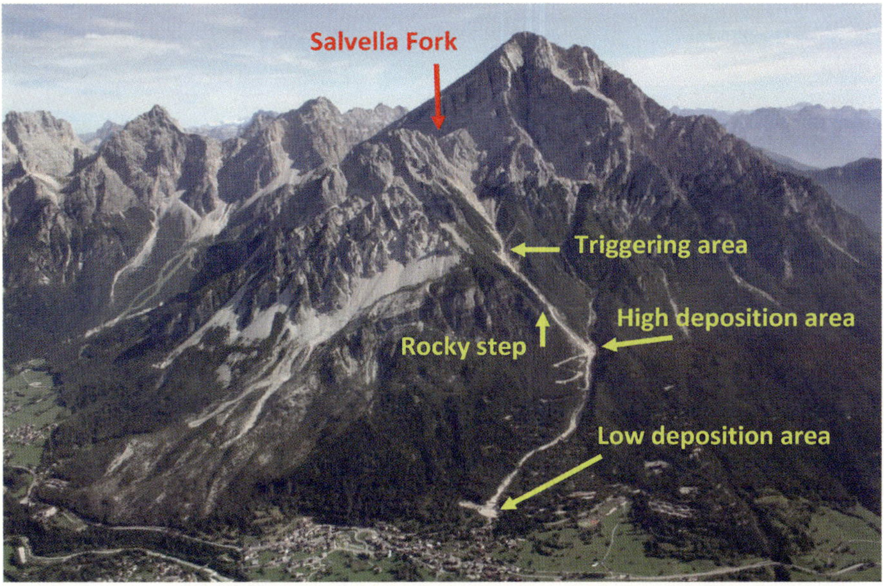

Fig. 2 Study area 'Ravina di Cancìa' with the two main hydrological basins. Source Ortofoto 2008 Regione Veneto mapped over a digital terrain model

Fig. 3 Details of the study site with loose gravel and surfacing rocks

time was set to 1 h with a sampling rate of 10 s. For the data collection, the following geodetic-grade GNSS receivers were employed: two Topcon HiPer Pro, a Leica Wild GNSS System 200 with SR299 antenna, and a Trimble 5700. A least squares free network adjustment was applied to each acquired data set, using all available independent baselines. The computations were executed using Topcon Tools' software adopting the 'single base' approach. Preliminary, loop closure analysis of the post-processed baselines showed all the time misclosures of a few millimetres, thus denoting the absence of any gross error in the observations. The resulting adjusted coordinates (N, E, h) of the ten control points are listed in Table 1. Here, the differences (ΔE, ΔN, Δh) have been computed with respect to the first survey epoch. Adjustment of the geodetic network yielded coordinate standard deviations at centimetre level.

3.2 Analysis of Displacements with the Bayesian Method

The Bayesian method applied to test significance over differences—in our case difference in vertical point position between epochs—tests the likelihood of such difference given in the prior model of expected residuals (i.e., height differences due to error budget of GNSS) of the measure. The general formulation of Bayes theorem states that if A and B are two stochastic variables, scalars or vectors, with known probability density functions, $P(A)$ and $P(B)$, the relation between the joint probability densities $P(A, B) = P(B, A)$, the conditional probability densities $P(A|B)$ and $P(B|A)$, and the single probability densities $P(A)$ and $P(B)$ are defined as follows:

$$P(A,B) = P(A|B)P(B) = P(B|A)P(A) \tag{1}$$

which can be written as follows:

$$P(A|B) = \frac{P(B,A)}{P(B)} = \frac{P(A,B)}{P(B)} = \frac{P(B|A)P(A)}{P(B)} \tag{2}$$

resulting in Bayes theorem:

$$P(A|B) = \frac{P(B|A)P(A)}{P(B)} \tag{3}$$

In other words, the likelihood that the difference is explained by the error model. On a simplified schema, if A is the accuracy in terms of expected normal distribution of the differences due to measuring errors and B is the real difference, we test the likelihood that B if significantly above the expected difference due expected error frequency distribution.

Table 1 Coordinates of the points in the GNSS network with their residual error in terms of standard deviation after differential correction (Ep. = Epoch)

	Ep.	E (m)	N (m)	h (m)	σ_E (m)	σ_N (m)	σ_h (m)	ΔE (m)	ΔN (m)	Δh (m)
1	1	287,273.634	5,145,139.688	990.154	0.03	0.03	0.08			
	2	287,273.627	5,145,139.684	990.145	0.05	0.05	0.10	0.007	0.004	0.009
	3	287,273.619	5,145,139.678	990.144	0.03	0.05	0.06	0.008	0.006	0.010
	4	287,273.614	5,145,139.675	990.145	0.04	0.05	0.06	0.005	0.003	0.009
2	1	286,220.336	5,146,347.164	968.639	0.03	0.04	0.10			
	2	286,220.332	5,146,347.159	968.625	0.03	0.04	0.11	0.004	0.005	0.014
	3	286,220.324	5,146,347.152	968.628	0.02	0.02	0.04	0.008	0.007	0.011
	4	286,220.318	5,146,347.145	968.624	0.03	0.04	0.10	0.006	0.007	0.015
3	1	287,771.711	5,146,198.934	1,279.527	0.04	0.04	0.09			
	2	287,771.705	5,146,198.930	1,279.517	0.03	0.03	0.08	0.006	0.004	0.010
	3	287,771.701	5,146,198.924	1,279.512	0.05	0.02	0.08	0.004	0.006	0.015
	4	287,771.694	5,146,198.921	1,279.514	0.03	0.05	0.07	0.007	0.003	0.013
4	1	289,574.834	5,147,974.169	3,164.527	0.02	0.02	0.11			
	2	289,574.830	5,147,974.161	3,164.441	0.02	0.04	0.04	0.004	0.008	0.086
	3	289,574.826	5,147,974.153	3,164.449	0.02	0.05	0.07	0.004	0.008	0.078
	4	289,574.821	5,147,974.150	3,164.429	0.02	0.05	0.07	0.005	0.003	0.098
5	1	288,398.021	5,147,007.454	1,746.699	0.04	0.03	0.08			
	2	288,398.014	5,147,007.431	1,746.600	0.04	0.02	0.08	0.007	0.023	0.099
	3	288,398.008	5,147,007.417	1,746.599	0.04	0.05	0.05	0.006	0.014	0.100
	4	288,397.991	5,147,007.387	1,746.627	0.05	0.02	0.04	0.017	0.03	0.072
6	1	288,395.084	5,146,895.914	1,674.800	0.03	0.05	0.09			
	2	288,395.061	5,146,895.896	1,674.715	0.02	0.03	0.11	0.023	0.018	0.085
	3	288,395.055	5,146,895.885	1,674.724	0.05	0.02	0.04	0.006	0.011	0.076
	4	288,395.033	5,146,895.879	1,674.722	0.05	0.05	0.11	0.022	0.006	0.078

(continued)

Table 1 (continued)

	Ep.	E (m)	N (m)	h (m)	σ_E (m)	σ_N (m)	σ_h (m)	ΔE (m)	ΔN (m)	Δh (m)
7	1	288,466.157	5,146,945.457	1,740.800	0.02	0.03	0.10			
	2	288,466.134	5,146,945.441	1,740.703	0.03	0.03	0.09	0.023	0.016	0.097
	3	288,466.112	5,146,945.412	1,740.709	0.02	0.03	0.06	0.022	0.029	0.091
	4	288,466.099	5,146,945.401	1,740.711	0.02	0.05	0.04	0.013	0.011	0.089
8	1	288,749.217	5,147,380.425	2,042.012	0.05	0.04	0.03			
	2	288,749.202	5,147,380.380	2,041.859	0.04	0.05	0.03	0.015	0.045	0.153
	3	288,749.154	5,147,380.346	2,041.872	0.02	0.02	0.02	0.048	0.034	0.140
	4	288,749.107	5,147,380.307	2,041.887	0.03	0.03	0.04	0.047	0.039	0.125
9	1	288,649.394	5,147,552.146	2,107.801	0.04	0.05	0.04			
	2	288,649.354	5,147,552.116	2,107.681	0.03	0.03	0.04	0.040	0.030	0.120
	3	288,649.339	5,147,552.094	2,107.652	0.04	0.04	0.05	0.015	0.022	0.149
	4	288,649.312	5,147,552.056	2,107.615	0.04	0.04	0.02	0.027	0.038	0.186
10	1	288,620.624	5,147,340.959	1,959.589	0.03	0.02	0.03			
	2	288,620.575	5,147,340.942	1,959.480	0.02	0.04	0.05	0.049	0.017	0.109
	3	288,620.565	5,147,340.918	1,959.407	0.04	0.04	0.03	0.01	0.024	0.182
	4	288,620.546	5,147,340.871	1,959.443	0.05	0.03	0.04	0.019	0.047	0.146

Points 5–7 and 8–10 are, respectively, the lower and upper group of points near the landslide (Fig. 1)

The Bayesian approach allows identifying in advance the areas on the ground with statistically significant shifts. A drawback of traditional statistical methods is that they work well only when point displacements between different survey epochs are sufficiently large compared to the standard deviations of related coordinates. In such cases, coordinate differences of some points can be marked as potential displacements by the classical methods. Bayesian analysis can help to better discriminate these 'ambiguities' (see Sacerdote et al. (2010)). Point shifts computed with respect to the first measurement epochs had the same order of magnitude as the residuals of the corresponding coordinates.

In order to evaluate whether the differences in vertical coordinates, with respect to the values at epoch 1, are more likely to be caused by a movement then to be due to random measurement errors, the Bayesian approach is more robust (see Betti et al. (2011)). The test was limited to the one-dimensional case since the results of classical analysis did not show any doubt on the horizontal components of the geodetic network points. Thus, for each point $P_{j,}$ the shifts Δh between different measurement campaigns were taken into account:

$$(\Delta h)_{P_j} = (h_i - h_1)_{P_j} \quad \text{with} \quad i = 2,\ldots,6 \quad \text{and} \quad j = 1,\ldots,5 \tag{4}$$

In Eq. (4), h_i denotes the adjusted height of point P_j at surveying epoch t_i, while h_1 is the adjusted height of the same point at the first measurement epoch t_1, considered as reference value. Assuming that the shifts Δh have a normal distribution with unknown mean δh and known variance σ_h^2 (computed from network adjustment), for each control point P_j the shift Δh can be written as:

$$\Delta h = h_i - h_1 = \delta h + \sigma_h \tag{5}$$

The mean δh is, in turn, a random variable following a normal distribution with mean μ and variance σ_0^2 which represents, in this analysis, the prior distribution of the Bayesian statistical inference. The parameters μ and σ_0^2 are the prior information whose values have to be somehow set in advance.

Since the points 5 through 10 were placed close to the landslide area, it is reasonable to assume that the vertical displacements can be zero or negative (decrease in altitude). Therefore, considering a properly oriented axis, the following additional a priori constraint has been set: $\delta h \geq 0$.

Considering the observables Δh as dependant on parameter δh, the Bayes formula becomes:

$$f(\delta h|\Delta h) = \frac{f(\Delta h|\delta h) \cdot f(\delta h)}{\int_{-\infty}^{+\infty} f(\Delta h|\delta h) \cdot f(\delta h) \cdot \delta h} \tag{6}$$

All the terms on the right of equation (above) can be explicitly calculated. The function $f(\delta h)$ in this analysis denotes a priori probability distribution of parameters δh. This distribution follows a modified version of a normal distribution: along the negative semi-axis, it is null, being the probability of the interval $[-\infty, 0]$ all concentrated in the origin, i.e. $P_0 \equiv P\{\delta h \leq 0\}$. Given this constraint, the probability distribution of δh becomes

$$f(\delta h) = P_0\delta(\delta h) + \frac{\vartheta(\delta h)}{\sigma_0\sqrt{2\pi}} \cdot \mathrm{e}^{-(\delta h-\mu)^2/2\sigma_0^2} \tag{7}$$

where (δh) is the unit step function (or Heaviside step function):

$$\vartheta(\delta h) = \begin{cases} 0 & \text{for} \quad \delta h < 0 \\ 1 & \text{for} \quad \delta h \geq 0 \end{cases} \tag{8}$$

and $\delta(\delta h)$ is the delta of Dirac function which is a generalized distribution that is zero everywhere except at zero.

The value of P_0 can be calculated by considering the normalization condition applied to the distribution probability $f(\delta h)$. Indeed, from Eq. (9):

$$P_0\delta(\delta h) + \int_0^{+\infty} f(\delta h) \cdot \mathrm{d}(\delta h) = 1 \tag{9}$$

it follows that

$$P_0 = \int_{-\infty}^{0} f(\delta h) \cdot \mathrm{d}(\delta h) = \frac{1}{\sigma_0\sqrt{2\pi}} \cdot \int_{-\infty}^{0} \mathrm{e}^{-(\delta h-\mu)^2/2\sigma_0^2} \cdot \mathrm{d}(\delta h) \tag{10}$$

The integral on the right side of (10) can be solved using the error function (Zwillinger 2012) whose values are available in specific tables:

$$\mathrm{erf}(z) = \int_{-\infty}^{z} \frac{1}{\sqrt{2\pi}} \mathrm{e}^{-t^2/2} \cdot \mathrm{d}t = \frac{1}{2} + \int_{0}^{z} \frac{1}{\sqrt{2\pi}} \mathrm{e}^{-t^2/2} \cdot \mathrm{d}t \tag{11}$$

This way, after a variable change, Eq. (7) becomes:

$$P_0 = \int_{-\infty}^{-\mu/\sigma_0} \frac{1}{\sqrt{2\pi}} \mathrm{e}^{-t^2/2} \cdot \mathrm{d}t = \mathrm{erf}\left(-\frac{\mu}{\sigma_0}\right) \tag{12}$$

In Eq. (4), the function $f(\Delta h|\delta h)$ can be regarded as the likelihood function $L(\Delta h|\delta h)$ of variable Δh:

$$L(\Delta h|\delta h) = \frac{1}{\sigma_h\sqrt{2\pi}} \cdot \mathrm{e}^{-(\Delta h-\delta h)^2/2\sigma_h^2} \tag{13}$$

The denominator of (6) is a normalization constant which can be numerically estimated. After some mathematical steps, the following formula is obtained:

$$K = A + B \tag{14}$$

with

$$A = \frac{P_0}{\sqrt{2\pi}\sigma_h} \mathrm{e}^{-\Delta h^2/2\sigma_h^2} \tag{15}$$

and

$$B = \frac{\bar{\sigma}\mathrm{e}^{\left(\frac{1^2}{2}\left[\frac{\Delta h^2}{\sigma_0^2} - \frac{m^2}{\bar{\sigma}^2}\right]\right)}}{\sqrt{2\pi}\sigma_h\sigma_0} \cdot \left[1 - \mathrm{erf}\left(-\frac{m}{\bar{\sigma}}\right)\right] \tag{16}$$

Being all terms in Eq. (6) defined in explicit form, the Bayes formula can be now numerically evaluated as follows:

$$\begin{aligned} f(\delta h|\Delta h) &= \frac{f(\Delta h|\delta h) \cdot f(\delta h)}{K} \\ &= \frac{\frac{1}{\sigma_h\sqrt{2\pi}} \cdot \mathrm{e}^{-(\Delta h - \delta h)^2/2\sigma_h^2} \cdot \left(P_0\delta(\delta h) + \frac{\vartheta(\delta h)}{\sigma_0\sqrt{2\pi}} \cdot \mathrm{e}^{-(\delta h - \mu)^2/2\sigma_0^2}\right)}{A + B} \end{aligned} \tag{17}$$

The benefit of using the two quantities A and B becomes clear by evaluating the probability that significant ($\delta h \geq 0$) or not significant ($\delta h = 0$) vertical displacements have occurred between 2011 and 2013. Indeed, this operation is turned into the calculation of the following simple ratios:

$$P(\delta h > 0|\Delta h) = \int_{-\infty}^{+\infty} P(\delta h|\Delta h) \cdot \mathrm{d}(\delta h) = \frac{B}{A + B} \tag{18}$$

$$P(\delta h = 0|\Delta h) = \frac{1}{A + B} \cdot \frac{P_0}{\sqrt{2\pi}\sigma_h} \cdot \mathrm{e}^{-\Delta h^2/2\sigma_h^2} = \frac{A}{A + B} \tag{19}$$

The significance analysis of displacements through the Bayesian approach is thus reduced to a comparison between the two quantities (18) and (19). A probabilistic analysis can be therefore performed instead of classical statistical testing. The result of the comparison allows to assess which of the two alternatives (significant or not significant shift) is more likely to be occurred.

Table 2 Criteria threshold for the Bayesian statistical analysis

$P(\delta h > 0 \vert \Delta h)$	Interpretation
<0.475	Point displacement is not significant → $P(\delta h = 0 \vert \Delta h) > 0.525$
>0.525	Point displacement is significant → $P(\delta h = 0 \vert \Delta h) < 0.475$
0.475 ÷ 0.525	No assessment can be made about the significance of the displacement

While in classical statistical analysis a decision rule based on a confidence level (e.g. $\alpha = 5\ \%$) in the Bayesian statistical analysis, a different approach was adopted, as shown in Table 2.

Three tests were then carried out with different settings for the prior values of parameters μ and σ_0. For each test, the probabilities $P(\delta h > 0|\Delta h)$ were calculated, assuming as reference for the comparisons the adjusted heights of the network points derived from the first measurement epoch. Although the classical analysis had highlighted some 'ambiguities' just for points *P*4 and *P*5, the Bayesian approach was applied to all the control points.

Finally to compare with classical statistics, we applied a z-score comparison using the residuals of the corrections for the network:

$$Z = \frac{\Delta h}{\sqrt{\sigma_{v_1}^2 + \sigma_{v_2}^2}} \tag{20}$$

where $\sigma_{v_1}^2$ and $\sigma_{v_2}^2$ are known as they have been calculated by the least squares network adjustment at the two surveying epochs t_1 and t_2, and Δh is the difference in height. In order to statistically check the significance of the network point displacements, computed within the surveys, the null hypothesis tested was that no significant displacements occurred between two measurement epochs.

4 Results and Discussion

Test results are illustrated in Table 3, where three different parameters for Bayesian data *prior* for each point in the CN are tested (A–C), and a column with the z-test scores are reported. Values which are above the threshold, and therefore significant, are in boldface. In the third Bayesian test, the value of the parameter μ was set equal to the mean shift Δh for each point P_j. The data prior (μ, σ_0) were set according to accumulated experience and knowledge about the landslide, derived from previous surveys. The first four points are not reported as they resulted in very low values, as expected, and we will focus on the points near the landslide body.

We can see from the data in boldface that the upper group of points (point 8, 9 and 10) have proven to be significantly shifting downwards. With a classical statistical test, only two points, 8 and 9, and only in one case for each of the three compared height differences, show that the recorded values can be considered as

Table 3 Results of Bayesian analysis, respectively, with (*A*) μ = 0.040 m, σ_0 = 0.02 m, (*B*) μ = 0.050 m, σ_0 = 0.05 m and (*C*) μ = variable, σ_0 = 0.08 m

Point	Epoch	Δh (m)	*A*	*B*	*C*	*z*-score
5	$t_2 - t_1$	−0.092	0.175	0.169	0.150	1.029
	$t_3 - t_1$	−0.1	0.210	0.123	0.062	1.060
	$t_4 - t_1$	−0.1	0.260	0.174	0.101	1.118
6	$t_2 - t_1$	−0.083	0.231	0.210	0.112	0.781
	$t_3 - t_1$	−0.095	0.213	0.207	0.170	1.062
	$t_4 - t_1$	−0.071	0.201	0.145	0.101	0.753
7	$t_2 - t_1$	−0.076	0.198	0.125	0.115	0.760
	$t_3 - t_1$	−0.079	0.274	0.180	0.168	0.837
	$t_4 - t_1$	−0.075	0.396	0.299	0.239	0.663
8	$t_2 - t_1$	−0.104	0.143	0.137	0.094	1.217
	$t_3 - t_1$	−0.187	**1.062**	**0.969**	**0.871**	**1.982**
	$t_4 - t_1$	−0.101	0.450	0.428	0.415	1.071
9	$t_2 - t_1$	−0.168	**0.963**	**0.894**	**0.823**	**2.037**
	$t_3 - t_1$	−0.114	**0.590**	**0.589**	0.320	1.334
	$t_4 - t_1$	−0.138	**0.527**	0.352	0.150	1.673
10	$t_2 - t_1$	−0.136	**0.709**	**0.638**	**0.619**	1.592
	$t_3 - t_1$	−0.119	**0.618**	**0.585**	**0.537**	1.393
	$t_4 - t_1$	−0.154	0.295	0.204	0.124	1.722

genuine shifts and not as false positives due to residuals of the measures. The application of the Bayes approach gave results which included also point 10 as significant. It also included more than one epoch of survey for all three points as significant with respect to the first measure (t_1). Historically, it is known that there is a downward shift of that part of the basin, and our results are compatible with such information.

The variation of the data *prior* shows that the method is robust, giving coherent results and pushing to the same conclusions. Point 9 seems particularly sensible to the test, and we can see from Fig. 1 that it is the point with higher altitude of the group; this suggests a relationship with the position along the main axis of the basin. It is a hypothesis which can be tested by adding points to the network and applying this approach to a larger network, or to another one positioned differently.

The application of the Bayes approach can also be used inversely, to prove that a CN is stable over time, and therefore, that it can be used as control points for stationing other instruments which might be used for change detection and/or landslide monitoring using remote sensing techniques that were mentioned in the introduction. In our study case, it was used to define points with significant vertical downward shift.

Traditional methods work well when coordinate differences between survey epochs are large enough, with respect to the residuals from error budget of the network adjustment, to remove any doubts on the cause of such difference. The Bayesian approach is not to be considered as a substitute to classical statistical testing, but as an integration, which can reliably give a more in-depth information clearing doubts in the case of borderline values. This can be done without having to increase the sample size, i.e. more surveys, as in the case of classical statistics, thus decreasing costs. The Bayesian approach works well when the movements have a preferred directionality; this might not be the case in many applications, so the phenomenon has to be interpreted beforehand and then a decision taken on which approach will work best.

References

Barbarella, M., Fiani, M., & Lugli, A. (2013). Landslide monitoring using multitemporal terrestrial laser scanning for ground displacement analysis. *Geomatics, Natural Hazards and Risk*, 21 (2013). doi:10.1080/19475705.2013.863808.

Barbarella, M., Fiani, M., & Lugli, A. (2014). Multi-temporal terrestrial laser scanning survey of a landslide. In M. Scaioni (Ed.), *Modern technologies for landslide investigation and prediction* (pp. 89–121). Berlin, Heidelberg: Springer.

Betti, B., Biagi, L., Crespi, M., & Riguzzi, F. (1999). GPS sensitivity analysis applied to non-permanent deformation control networks. *Journal of Geodesy, 73*, 158–167.

Betti, B., Sansò, F., & Crespi, M. (2001). Deformation detection according to a Bayesian approach. In: B. Benciolini (Ed.), *IV Hotine-Marussi Symposium on Mathematical Geodesy, IAG Symposia*, (Vol. 122, pp. 83–88). Heidelberg: Springer. doi:10.1007/978-3-642-56677-6_12.

Betti, B., Cazzaniga, N. E., & Tornatore, V. (2011). Deformation assessment considering an a priori functional model in a bayesian framework. *Journal of Surveying Engineering, 137*, 113–119.

Bitelli, G., Dubbini, M., & Zanutta, A. (2004). Terrestrial laser scanning and digital photogrammetry techniques to monitor landslide bodies. *International Archives of the Photogrammetry, Remote Sensing and Spatial Information Sciences, 38*(7B), 246–251.

Borghi, A., Cannizzaro, L., & Vitti, A. (2012). Advanced techniques for discontinuity detection in GNSS coordinate time-series. an italian case study. In: S. Kenyon et al. (Eds.), *Geodesy for Planet Earth, IAG Symposia* (Vol. 136, pp. 627–634). Heidelberg: Springer. doi:10.1007/978-3-642-20338-1_77.

Cina, A., & Piras, M. (2014) Performance of low-cost GNSS receiver for landslides monitoring: test and results. *Geomatics, Natural Hazards and Risk*, p. 18. doi:10.1080/19475705.2014.889046.

Crosetto, M., Monserrat, O., Luzi, G., Cuevas-González, M., & Devanthéry, N. (2014). Discontinuous GBSAR deformation monitoring. *ISPRS Journal of Photogrammetry and Remote Sensing, 93*, 136–141.

Dowman, I. (2004). Integration of LiDAR and IFSAR for mapping. *The International Archives of The Photogrammetry, Remote Sensing and Spatial Information Sciences, 35*(2), 11.

Farina, P., Colombo, D., Fumagalli, A., Marks, F., & Moretti, S. (2006). Permanent scatterers for landslide investigations: outcomes from the ESA-SLAM project. *Engineering Geology, 88*(3–4), 200–217.

Forlani, G., Roncella, R., & Diotri, F. (2013). Production of high-resolution digital terrain models in mountain regions to support risk assessment. In: *Geomatics, Natural Hazards and Risk* (p. 19). doi:10.1080/19475705.2013.862746.

Frangioni, S., Bianchini, S., & Moretti, S. (2014). Landslide inventory updating by means of persistent scatterer interferometry (PSI): The Setta basin (Italy) case study. In: *Geomatics, Natural Hazards and Risk*. p. 20, doi:10.1080/19475705.2013.866985.

Gregoretti, C., & Dalla Fontana, G. (2008). The triggering of debris flow due to channel-bed failure in some alpine headwater basins of the dolomites: Analyses of critical runoff. *Hydrological Processes, 22*, 2248–2263.

Hofmann-Wellenhof, B., Lichtenegger, H., & Wasle, E. (2007). *GNSS—global navigation satellite systems: GPS, GLONASS, Galileo, and more* (p. 516). Heidelberg: Springer.

IGM (2014). Rete Geodetica Nazionale. http://www.igmi.org/geodetica/. Accessed July 10th 2014.

Jaboyedoff, M., Oppikofer, T., Abellán, A., Derron, M. H., Loye, A., Metzger, R., et al. (2012). Use of LIDAR in landslide investigations: A review. *Natural Hazards, 61*, 1–24.

Koch, K. (2007). *Introduction to Bayesian statistics*. Heidelberg: Springer.

Lingua, A., Piatti, D., & Rinaudo, F. (2007). Remote Monitoring of a landslide using an integration of GB-InSAR and LiDAR techniques. *The International Archives of the Photogrammetry, Remote Sensing and Spatial Information Sciences, 37*(B1), 361–366.

Manfré, L. A., Hirata, E., Silva, J. B., Shinohara, E. J., Giannotti, M. A., Larocca, A. P. C., et al. (2012). An analysis of geospatial technologies for risk and natural disaster management. *ISPRS International Journal of Geo-Information, 1*(3), 166–185.

Monserrat, O., Crosetto, M., & Luzi, G. (2014). A review of ground-based SAR interferometry for deformation measurement. *ISPRS Journal of Photogrammetry and Remote Sensing, 93*, 40–48.

Pirotti, F., Guarnieri, A., & Vettore, A. (2013). Vegetation filtering of waveform terrestrial laser scanner data for DTM production. *Applied Geomatics, 5*(4), 311–322.

Previtali, M., Barazzetti, L., & Scaioni, M. (2014). Accurate 3D surface measurement of mountain slopes through a fully automated imaged-based technique. *Earth Science Informatics, 7*, 109–122.

Refice, A., Bovenga, F., Wasowski, J., & Guerriero, L. (2000). Use of InSAR data for landslide monitoring: a case study from southern Italy. In: Proceedings of the IGARSS 2000, Honolulu, 24–28 July 2000, (Vol. 6, pp. 2504–2506). doi:10.1109/IGARSS.2000.859621.

Remondino, F., Spera, M. G., Nocerino, E., Menna, F., & Nez, F. (2014). State of the art in high density image matching. *The Photogrammetric Record, 29*(146), 144–166.

Sacerdote, F., Cazzaniga, N. E., & Tornatore, V. (2010). Some considerations on significance analysis for deformation detection via frequentist and Bayesian tests. *Journal of Geodesy, 84*, 233–242.

Scaioni, M., Feng, T., Barazzetti, L., Previtali, M., Lu, P., & Qiao, G., et al. (2014a). Some applications of 2-D and 3-D photogrammetry during laboratory experiments for hydrogeological risk assessment. *Geomatics, Natural Hazards and Risk*, 24 (2014). doi:10.1080/19475705.2014.885090.

Scaioni, M., Feng, T., Lu, P., Qiao, G., Tong, X., Li, R., et al. (2014b). Close-range photogrammetric techniques for deformation measurement: Applications to landslides. In M. Scaioni (Ed.), *Modern Technologies for Landslide Investigation and Prediction* (pp. 13–41). Berlin, Heidelberg: Springer.

Wang, G., & Soler, T. (2012). OPUS for horizontal subcentimeter-accuracy landslide monitoring: case study in the Puerto rico and Virgin islands region. *Journal of Surveying Engineering, 138* (3), 143–153.

Ye, X., Kaufmann, H., & Guo, X. F. (2004). Landslide monitoring in the three gorges area using D-INSAR and corner reflectors. *Photogrammetric Engineering & Remote Sensing, 70*, 1167–1172.

Zwillinger, D. (2012). *CRC Standard Mathematical Tables*.

Part II
Geophysical and Geotechnical Methods

Analysis of Microseismic Activity Within Unstable Rock Slopes

Diego Arosio, Laura Longoni, Monica Papini and Luigi Zanzi

Abstract This chapter illustrates the concept of passive seismics as a method for monitoring the propagation of cracks within a rock mass as a result of load stress or water freezing in view of the use of this technique for rockfall early warning. The methodology is still far from being a standard and consolidated technique. The research is making progress, but just a few real case studies are documented. They are shortly overviewed in the introduction. Then, an interesting field test where crack propagation was artificially triggered up to full rock detachment, while a small sensor network was active, is discussed to show the existence and the characteristics of precursory signals. It follows the illustration of the microseismic monitoring methodology through the description of the Mt. San Martino (Lecco, Italy) sensor network and the discussion of the preliminary results obtained during the initial months of activity. Apparently, the preliminary results show some correlation with rainfalls, but not with temperature. Microseismic spectra are mainly concentrated in the first 100 Hz. This probably means that the hypocentre distances from the sensors are quite longer than 10 m. Electromagnetic interferences are also observed as mentioned by other authors who have analyzed data sets from other microseismic networks installed in mountain regions. They are automatically discriminated from significant signals by a classification software which works on the time/frequency properties of these events. Hypocenter localization and clustering analysis of the significant events are the planned near-future activities.

D. Arosio (✉) · L. Longoni · M. Papini · L. Zanzi
Department of Civil and Environmental Engineering, Politecnico di Milano,
Piazza Leonardo da Vinci 32, 20133 Milan, Italy
e-mail: diego.arosio@polimi.it

L. Longoni
e-mail: laura.longoni@polimi.it

M. Papini
e-mail: monica.papini@polimi.it

L. Zanzi
e-mail: luigi.zanzi@polimi.it

M. Scaioni (ed.), *Modern Technologies for Landslide Monitoring and Prediction*,
Springer Natural Hazards, DOI 10.1007/978-3-662-45931-7_7

Keywords Unstable rock slopes · Passive seismics · Microseismic sensor network · Data classification · Hypocenter localization · Rockfall

1 Introduction: Passive Seismics for Unstable Rock Slope Monitoring

Compared to other passive seismic applications such as earthquake monitoring or reservoir hydrofracturing monitoring, landslide and rockfall monitoring is at the very beginning. Earthquake monitoring has a very long tradition, the technology is consolidated, and the data interpretation is robust. Reservoir monitoring during hydrofracturing is a quite recent application, but remarkable advancements have been achieved in a very short time thanks to the enormous interest of the hydrocarbon industry. Besides, this application has one important advantage compared to landslides and rock slopes, i.e., the absence of environmental noise when the technique is applied with borehole sensor arrays. On the contrary, the limited funding of new initiatives in the civil protection sector and the peculiarities of the application of the passive seismic principle to unstable rock slope monitoring explain why this approach is still preliminary, why only a few examples of running monitoring projects exist, and why the expertise in the interpretation and use of rock slope seismic noise are still so poor.

One of the first examples which are well documented is the case of the Randa rockslide in the Swiss Alps (Spillman et al. 2007). The monitoring network was installed by ETH (Zurich) and has been active from 2002 to 2004. It consisted of twelve 3-axial sensors partly installed in boreholes and partly on the surface. The recorded microseismic activity was used to characterize the slope body. Hypocenter localization was performed by using a 3D velocity model created with a tomographic experiment.

The Aknes rock slope (Norway) is another example of microseismic monitoring. The network was installed by NORSAR (Blikra 2008, 2012) and has been active since 2005. Eight 3-axial geophones were installed. Data are systematically analyzed to perform event detection and classification. No localization is performed because a 3D velocity model is not available. The final objective is to monitor the risk of a big rockfall which might produce a tsunami since the unstable rock slope is facing a fjord.

More recent are the cases of Matterhorn (Swiss-Italian Alps) and Sechilienne (French Alps) monitoring networks. Both have been active since 2007. The Matterhorn network was installed by CNR (Amitrano et al. 2010) at very high altitude (about 3,830 m) close to the Carrel hut. It consists of five 3-axial high-frequency geophones installed in holes (3–5 m deep) and three 3-axial low-frequency geophones installed on surface. Because of altitude, the fractured rock slope is highly affected by freeze–thaw cycles which can represent a major cause of crack development and propagation. The Sechilienne network was installed by the Fourier

University of Grenoble (Helmstetter and Garambois 2010) and consists of three seismological stations that work in combination with three arrays of vertical sensors (one linear with 21 sensors and two circular with six sensors each). The network has been recording both internal cracks and rockfalls. Since 2009, a camera has been also installed to observe major rockfalls in order to correlate images and microseismic data in the attempt to calibrate the seismic interpretation in terms of falling rock volumes. Interesting correlations of microseismic events and rainfalls have been observed and studied at this site.

Another interesting case study is represented by the Heumoes slope monitoring network in the Vorarlberg Alps (Walter et al. 2012). It was made operative in 2009 by installing 3 seismic arrays, and in 2011, the system recorded an event produced by a major rockfall occurring at a distance of 5 km from the network. Extremely interesting is the fact that some hours in advance, the system also recorded some precursory events, partly produced by minor rockfalls and partly by fracture propagation.

Another case where a monitoring network had the chance to record precursory microseismic signals prior to an important rockfall is the Mesnil-Val study (Normandie, France). The network was installed on a coastal chalk cliff and consisted of two vertical stations in holes drilled on the top of the cliff (10 m deep) and three horizontal stations in holes drilled on the cliff face (6 m deep). Each station is composed by a geophone and an accelerometer to monitor a wide frequency up to 10 kHz. The network has been running from 2002 to 2004, and a major rockfall of about 1,000–2,000 m^3 occurred during the active period. Precursory signals were recorded since 15 h prior to the rockfall (Senfaute et al. 2009).

Inspired by these examples, we have been studying the problem of monitoring a limestone rock slope (Mt. San Martino) that is threatening the city of Lecco (northern Italy) where a campus of Politecnico di Milano University is located. Before installing a microseismic network and a meteorological station, we carried out some preliminary analysis. First, we set up some field tests on the typical limestones of the Prealpine region to explore its seismic properties such as velocity and absorption in order to predict the expected frequency range according to the sensor distance from the hypocenter (Arosio et al. 2009a). In that occasion, a comparison of the sensitivity of the different available technologies (MEMS accelerometers, piezoelectric transducers, geophones) was also accomplished. After that, we planned some laboratory and field tests aimed at studying the time–frequency characteristics of microseismic signals generated by crack propagation in the specific limestone rock which is typical of the Prealpine group of our selected rock slope. The objective of laboratory test was to reproduce the microseismic effects of freeze–thaw cycles on small rock samples (Arosio et al. 2013, 2014). The objective of the field test was to reproduce a gradual detachment of a small rock mass while monitoring the precursory microseismic signals with some piezo-accelerometers. The results of this test are summarized in the next section of this chapter, while Sects. 3 and 4, respectively, illustrate the microseismic network installed on the San Martino rock slope and the preliminary data analysis performed on the initial monitoring period.

2 Induced Crack Propagation Test

The objective of this test was to capture the signature of a typical crack that propagates in the rock face as a result of load stress or water freezing. The test was operated by soliciting the detachment of a small rock block (approximately 0.5 m^3) at the base of a limestone rock tower affected by a set of remarkable vertical fractures. This site was selected because of the proximity to the rock slope where the permanent monitoring network has been installed, sharing the same geological history. Thus, the results of this test should represent a good reference for the data analysis of the network records. Block failure was induced by means of a rectangular hydraulic jack inserted in an existing fracture; oil was manually pumped into the jack until complete block failure (actually, the block was completely detached but did not fall because it was held in place with a fastening system due to safety reasons). Three uniaxial piezoelectric accelerometers (Wilcoxon 799LF, sensitivity 500 mV/g, band 0.2–2,500 Hz) were deployed onto the rock surface at increasing distances from the hydraulic jack (Fig. 1) to collect microseismic signals generated by the forced fracture propagation. A 24-bit acquisition board continuously recorded signals with a sampling frequency of 5,120 Hz according to the band of the used transducers and in order to keep storage space to a minimum. Pressures as high as 0.45 MPa were reached and it took nearly 325 s to reach the block collapse. Sixty-nine events were detected during the experiment by the closest sensor, twenty-six events by the middle one, and just seven by the furthest sensor. More in

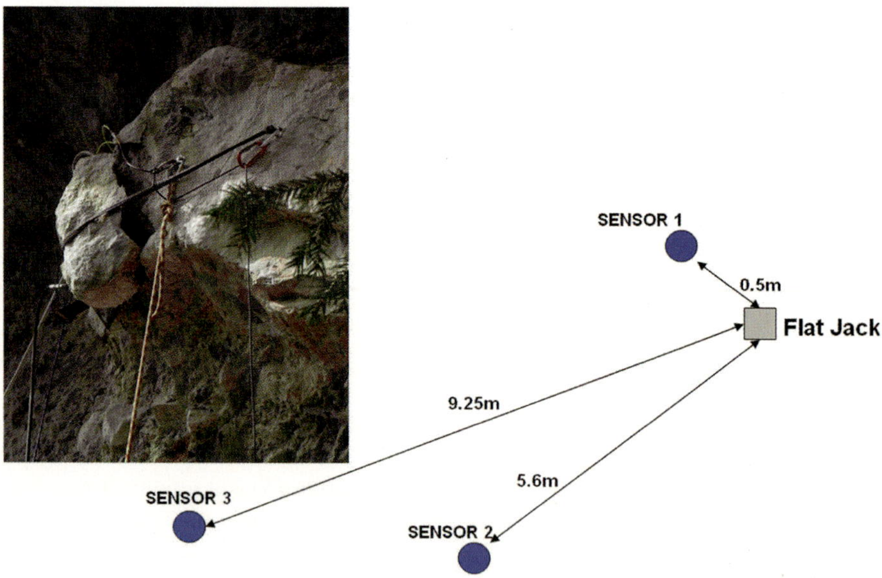

Fig. 1 Sensor positions and distances from the crack armed with the flat jack. The picture shows the rock block after full detachment and the ropes used to control the rockfall

detail, the furthest accelerometer was able to record meaningful signals just during the final burst sequence, when the strongest events were generated immediately before the ultimate collapse. As the two farthest accelerometers had similar orientation with respect to the source position, we also tried to estimate elastic wave attenuation by comparing the amplitude of the events detected by both sensors. We found an attenuation factor as high as 1.10 dB/m, which is in good agreement with other previous tests executed in another site but on the same type of limestone (Arosio et al. 2009a). By analyzing the events collected by the closest sensor, we observed wide bandwidth events with energy equally distributed in the whole frequency band of the recording system (Fig. 2 at the top). Thus, we cannot exclude that higher frequencies were also generated by the cracking phenomenon. However, a dramatic decrease of bandwidth is observed as soon as we move a few meters far

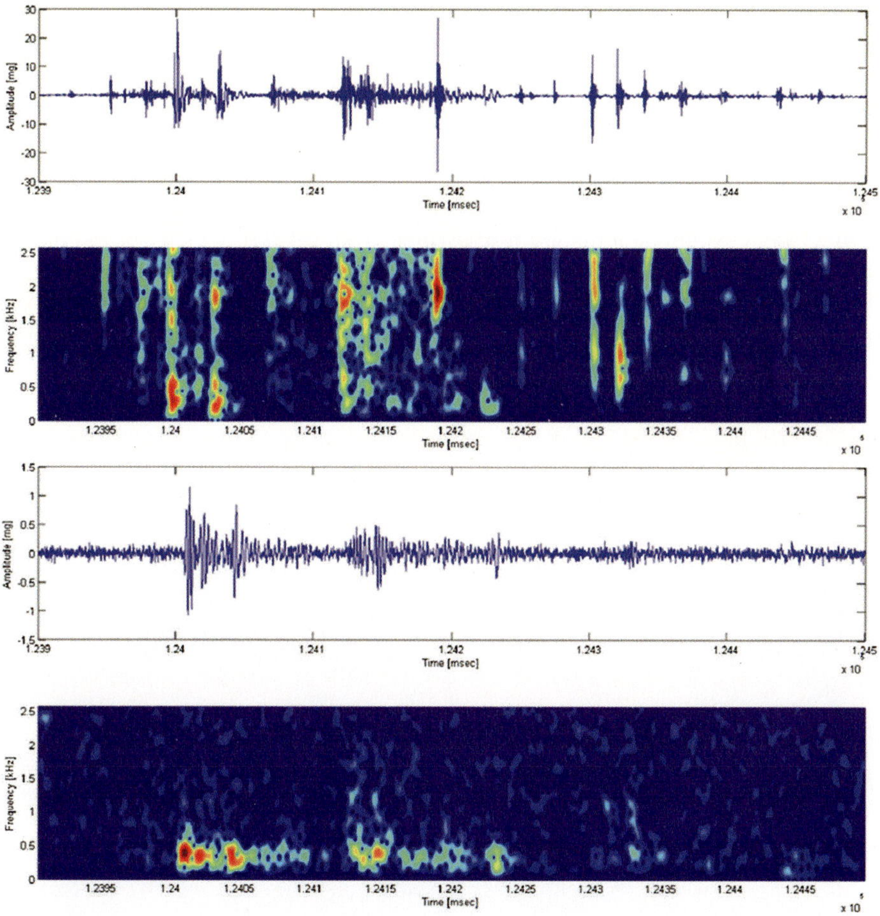

Fig. 2 Time–frequency representation of the final burst sequence recorded by the closest sensor (at the *top*) and the farthest sensor (at the *bottom*), respectively

from the crack position. Figure 2 (bottom) suggests that because of attenuation, an effective bandwidth of 500 Hz is sufficient to record most of the energy that travels at a distance of 10 m or more from the active crack.

In summary, this test indicated that a collapse is preannounced by an intense sequence of microseismic events with amplitude increasing as we approach the rockfall. Frequency spectra and attenuations observed at the different sensors suggest that a sampling frequency of 1 or 2 kHz is sufficient for a network of sensors with inter-distances in the order of a few tens of meters. Finally, the signatures of the events observed in this test will be a good reference for guiding the interpretation of the great amount of events recorded by the Mt. San Martino monitoring network.

3 Mt. San Martino Monitoring Network

Mt. San Martino is located at the end of the Lecco branch of the Como Lake (northern Italy). It is part of the Prealpi Lombarde and threatens the city of Lecco with the 300-m-high unstable rock face (Fig. 3). A major event occurred in 1969 caused 7 fatalities with a rockfall of about 15,000 m^3. The area that is currently monitored by the new microseismic network is close to the main scarp of the 1969 rockfall.

According to our preliminary test (Arosio et al. 2009a) carried out to compare the performances of different sensor technologies on the limestone rock of this Prealpi group, traditional sensors (i.e., geophones and piezoelectric transducers) are still remarkably superior to the low-cost MEMS accelerometers. Thus, we decided to install geophone sensors which can ensure in the required bandwidth (from a few Hz to 500 Hz) sensitivity comparable to that of piezo-transducers at a lower price.

Fig. 3 Mt. San Martino rock face threatening the city of Lecco (northern Italy)

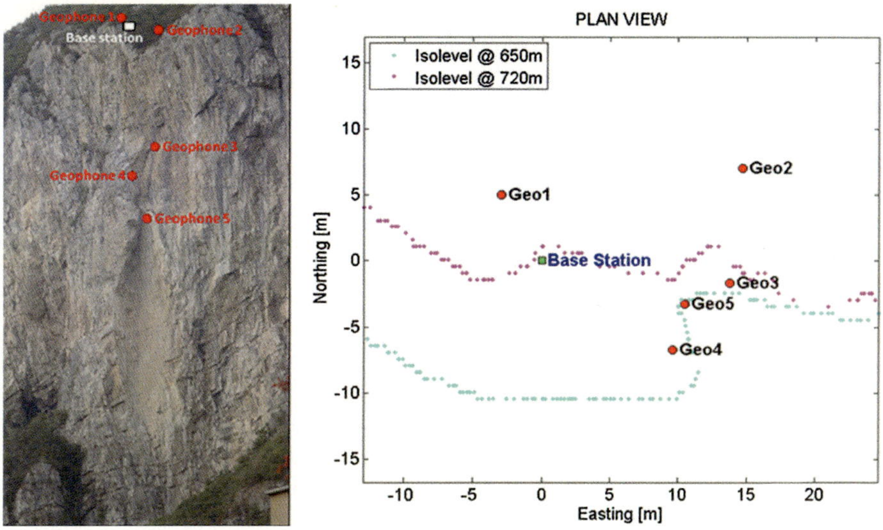

Fig. 4 Geophone positions on the rock slope (*left*) and plan view of the network geometry

The network consists of five 3-axial 28 Hz geophones plus a rain gauge and two temperature sensors (one for air temperature and one placed in a superficial fracture to check for freezing–thaw cycles).

Apart from three geophones installed in shallow holes onto the rock face, two geophones were deployed in vertical holes, 5 and 10 m deep, respectively, on the plateau above the rock face. The borehole sensors are important because they can provide data with higher signal-to-noise ratio (or sense weaker events) thanks to a quieter environment and improve the accuracy of source localization because of a better recording geometry (Arosio 2010). The network is extended more vertically than horizontally. The geophone position is illustrated in detail in Fig. 4. The elevation range, from the lowest to the highest geophone, approximately goes from 640 m to 720 m a.s.l.

The acquisition unit is equipped with an 18-channel 24-bit A/D conversion board and with a GPS receiver for correct network timing. Microseismic events are recorded when a selected short-time average/long-time average (STA/LTA) threshold is overcome by at least a single channel. Once a trigger occurs, the recorded event has a min–max duration range of 5–20 s with 2 s of pre- and post-trigger. Thus, multiple or prolonged events can be saved in a single records provided that they are not longer than 20 s. Sampling frequency for the microseismic events is 1 kHz, while meteorological parameters and the background seismic noise level are regularly collected at 10 s intervals. Other important parameters such as gain levels, threshold level, and filters were experimentally optimized during a preliminary calibration period in order to maximize the sensitivity of the network while trying to prevent the recording of non-significant noise events.

Power supply is provided by solar panels, and remote communication is granted by a dedicated radio link. Data are stored on a 2 GB CF card and daily transmitted to a central station located at the Lecco Campus of Politecnico di Milano.

4 Preliminary Data Analysis

The sensor network has been operating for more than one year, but data collection suffered from several power interruptions because of significant snowfall winter 2013–2014. In the first 7 months, the network recorded more than 5,000 events. We explored any possible correlation of these events with meteorological parameters. It seems that no correlation exists with temperature, but it is important to point out that the winter temperature was not particularly rigid, and according to temperature recorded by the sensor installed in a fracture, no freeze–thaw cycles occurred. Instead, a moderate correlation was observed with rainfalls. Figure 5 illustrates the cumulative number of events of the initial 7 months compared with the cumulative rainfall. Group of microseismic events is often observed after significant rainfalls but with a delay of a few days. This delay is longer in winter (4 or 5 days) and tends to reduce to 1 or 2 days toward summer. Interestingly, this sort of delayed correlation with rainfalls was also observed by Helmstetter and Garambois (2010).

A detailed analysis of time and frequency properties of our records shows that the events can be roughly classified with four types of signals: (a) single microseismic events, (b) multiple microseismic events, (c) spikes, and (d) noise, where only type (a) and (b) present properties that are quite similar to the expected type of events observed during the induced crack propagation test (Fig. 6). The amplitude spectra show that no significant energy is observed above 100 Hz, probably

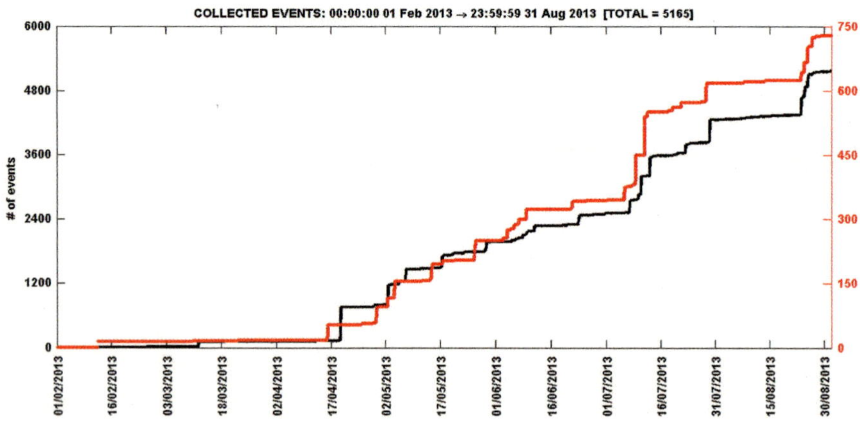

Fig. 5 Cumulative number of events in the initial 7 months of sensor network activity (*black line*) compared with cumulative rainfall (*red line*)

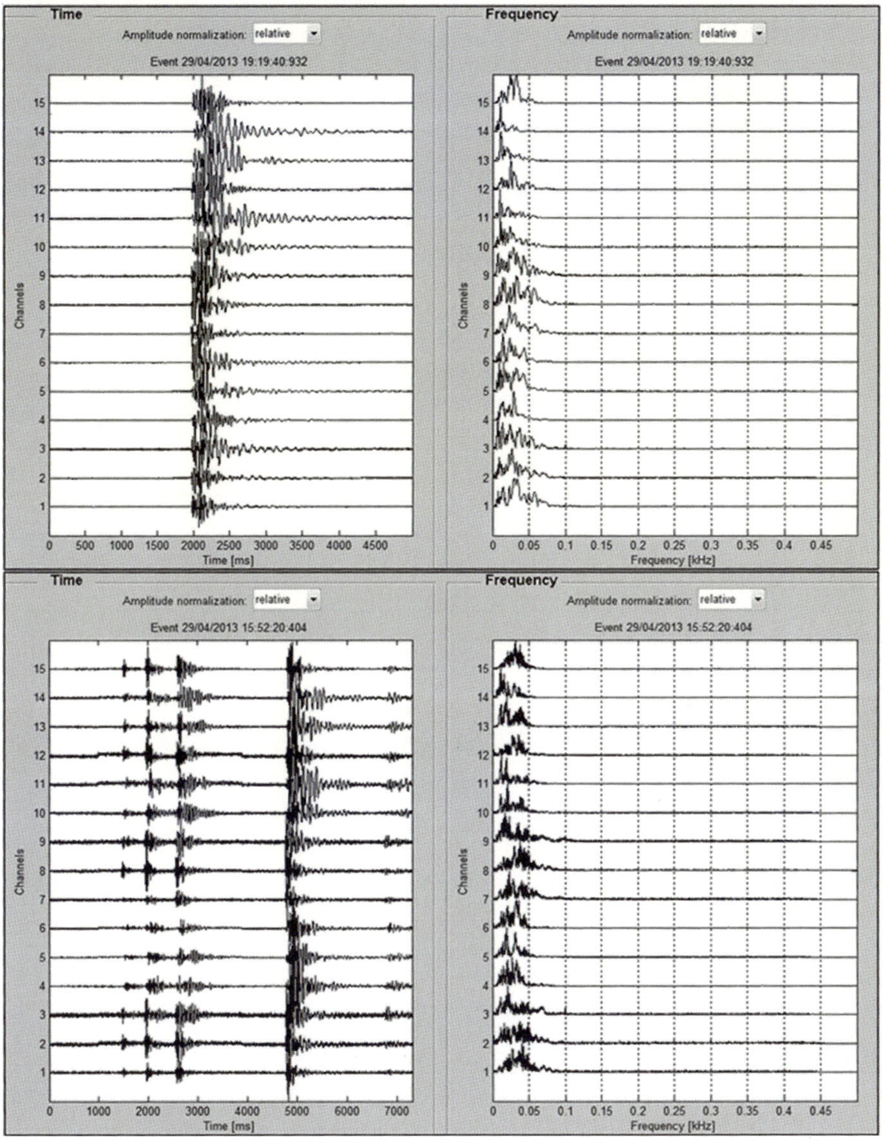

Fig. 6 Examples of time and frequency representation of significant single (at the *top*) and multiple (at the *bottom*) microseismic events

meaning that the hypocenters are quite farther than 10 m from the closest sensor. Figure 7 shows the typical signature of those events that we classify as spikes. The signature is unusually impulsive. The spectra span over the total available bandwidth (500 Hz) which probably means that the spectra were actually larger and were limited by the anti-alias filter of the acquisition board. It is hard to believe that

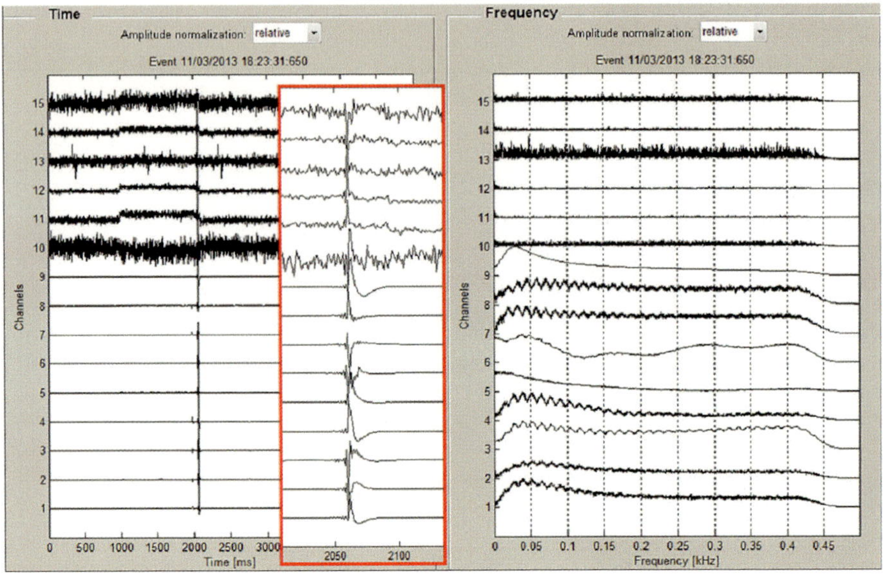

Fig. 7 Example of time and frequency representation of a spike event. In the *red box,* the timescale has been expanded around the spike

these events could be produced by crack propagation or other mechanical phenomena (e.g., micro-rockfalls). It is much more probable that these events result from electromagnetic interferences whose nature is still to be understood. The railway runs at the base of the rock face (i.e., at a distance of about 500 m), but no direct correlation has been demonstrated with the train timetable. Spikes occur during both day and night, while no trains are planned during the night and spikes are often distributed in groups that last for time intervals much longer than a train passage interval below the rock slope. Similarly, we did not find a correlation of spikes with local weather storms, but maybe the geographic area of our storm analysis should be expanded. A review of this correlation analysis is planned in order to consider a much larger geographic area. Thus, currently, we do not have any convincing explanation for these type (c) events. Again, it is interesting to note that a similar type of events was also noted by Spillmann et al. (2007). Finally, events of type (d), classified as noise, are records where an incoherent noisy signal lasts for the whole duration of the record on all the channels and occasionally hardly overtook the triggering threshold on one of the 15 channels. These noise events are possible because a very low STA/LTA threshold was selected to prevent any loss of significant events.

Given the large number of events (more than 5,000 in 7 months), we developed a software for automatic selection of significant events based on time and frequency characteristics of collected signals as well as on the number of sensors involved. As an example, we wanted to scrutinize from the 5,165 events recorded in the initial 7 months of monitoring all the events that excited more than three sensors simultaneously. A data set of 72 events was extracted. Among these, the software distinguished 20 spikes, 18 noise events, and 32 significant events (i.e., events of type (a) and (b)). Provided the SNR of the 32 selected events is high enough to allow a reliable traveltime picking, these events are those that could be submitted to a localization process to create a hypocenter distribution map from which to derive insights about the rock face internal fracturing process. As a matter of fact, four sensors represent the minimum setup needed to perform a 3D localization based on time difference of arrival (TDOA), see Hardy (2003) and Herath and Pathirana (2013). Figure 8 shows one of the 32 selected events as it was recorded by the five geophones. Raw data without any filtering or other processing are shown except local trace normalization. Data quality is good, and a fairly reliable traveltime picking can be performed on all the components of these geophones. From a rough analysis of the delays, it seems that the event reached the geophones in the following order: no. 4, no. 5, no. 3, no. 2, no. 1 so that the hypocenter is expected to be closer to the geophones installed on the rock face (no. 3, no. 4 and no. 5) rather than to the borehole geophones (no. 1 and no. 2). Actually, a critical issue in determining the hypocenter of the event is the velocity of the medium. Assuming a constant velocity medium, TDOA from two sensors confines the source location to a 3D hyperboloid and the intersection of three or more independent hyperboloids defines the hypocenter. Unfortunately, most authors who tried to perform 3D location of microseismic events generated by rock fracture propagation agree on saying that the

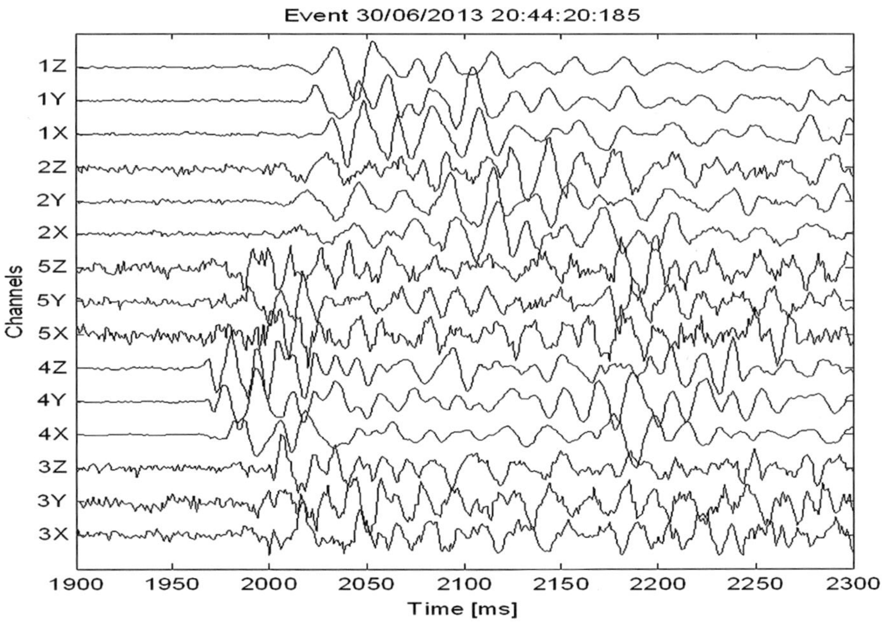

Fig. 8 Example of a single event detectable on all the components of the five 3-axial geophones

constant velocity assumption is too rough to obtain consistent results (Helmstetter and Garambois 2010; Spillmann et al. 2007). Thus, a more realistic velocity model is needed and the localization algorithm must incorporate a forward modelling software to calculate the signal trajectory through the fractured rock.

Another approach for 3D localization is the *coalescence* method (Drew et al. 2005) which helps handle data with bad SNR where traveltime picking becomes questionable, but which does not overcome the problem of the velocity field. This method consists of 3D focusing the recorded data in the supposed source position (Arosio et al. 2009b). Thus, a velocity field is still needed to apply the focusing operator. The hypocenter is found as the location where the highest amount of energy is collapsed by the algorithm. The method is widely applied to process microseismic data resulting from monitoring of reservoir hydrofracturing operations since the method can be applied automatically in order to get a real-time feedback for optimizing the proppant injection parameters. Compared to the TDOA algorithm where the result is exclusively determined by the traveltime delays, the coalescence method is also influenced by the energy observed by each sensor. Depending on the strategy that is applied to manage the energy received by the different sensors, the localization result may change. If no gain is applied, it is clear that the result will be mostly influenced by the sensors closer to the hypocentre, which are expected to receive much more energy than the farther receiver. Thus, energy in the coalescence method can work like a sort of weighting function applied to delay traveltimes. This is another possible advantage offered by the coalescence method compared to TDOA, although a sort of SNR-based weighting function could be also introduced in the TDOA process. In principle, a localization algorithm exclusively based on energy back-propagation analysis would be also applicable, provided that a reliable velocity field and an absorption field are assumed. As an example, by analyzing the energy collected by each geophone for the event illustrated in Fig. 8 (under the assumption of a constant absorption field), we found that the geophone closest to the hypocenter is still geophone no. 4 in full agreement with the TDOA indication, but the position for the other geophones do not match the order proposed by TDOA. Furthermore, the energy method also implies the necessity of neglecting the geophone coupling problem, i.e., the necessity to assume that the geophones are equally coupled or the necessity to calibrate the sensor network with specific experiments designed to estimate a coupling factor for each geophone. Given the difficulties associated with obtaining a good estimate of the absorption field and of the geophone coupling factor, it is hard to believe that a localization algorithm exclusively based on energy might compete with a TDOA or a coalescence method.

Currently, we do not have an opinion based on real data feedback between TDOA and coalescence methods because we still need to estimate a good velocity field for our unstable rock slope. To this aim, a tomographic experiment has been designed and will be performed before the next winter season.

Localization will be fundamental also for helping the classification of the significant events. As a matter of fact, among the 32 significant events of the initial 7 months, there might be events generated by a fracture propagation process but also events generated by a micro-rockfall (small stones hitting the rock face while falling down from the top of the rock face). Unfortunately, this occurrence is not improbable because the rock face, although quite vertical, is surmounted by a more gentle slope where rock debris accumulate and slide down especially during storms.

Thus, another priority for the near future is the classification of the significant events. To this aim, a database of the significant events is under construction. Any event is analyzed to extract a number of properties which include duration of the event, energy, SNR, bandwidth, spectral centroid, centroid variance, recording geophones, weather, and other parameters.

The goal is to complete this database as soon as a reliable velocity field is available with the localization of the events which excited more than four geophones and then to apply a clustering process to explore the possibility to get a clear discrimination at least between fracturing events and micro-rockfall events.

5 Conclusions

The way to an effective use of a microseismic monitoring sensor network for early warning of rockfall risk from an unstable rock slope is still quite long. According to our preliminary analysis of data collected on the Mt. San Martino rock face in Lecco (northern Italy), we can summarize the results as follows. Microseismic activity seems to have no correlation with temperature. However, this is a result which comes from a temperate winter during which according to our temperature sensors no freeze–thaw cycles occurred within the rock fractures. Instead, a moderate correlation exists with rainfalls although with a delay of some hours or a few days depending on the season (shorter delays in summer). No significant energy is observed above 100 Hz which probably means, according to our induced crack test, that the hypocenters are quite farther than 10 m from the closest sensor. Similarly to other microseismic monitoring networks installed on mountain regions, the Mt. San Martino network is also recording a great number of large bandwidth spikes, which we classified as electromagnetic interferences, whose nature is still under investigation.

Automatic discrimination software has been developed to extract significant events, and a database of significant event properties (e.g., duration, energy, SNR, bandwidth, spectral centroid, recording geophones, and weather) is under construction.

Several near-future activities are already planned: (a) a traveltime and attenuation tomographic experiment aimed at building a 3D velocity and absorption model, which according to most authors is absolutely needed to perform a consistent hypocenter localization of the events; (b) a classification process aimed at exploring how the significant events cluster according to their main properties and how the fracturing events can be discriminated from micro-rockfall events.

Acknowledgments The authors are grateful to Lecco Municipality and Province government that partially funded the development and installation of the microseismic monitoring system on Mt. San Martino rock face.

References

Amitrano, D., Arattano, M., Chiarle, M., Mortara, G., Occhiena, C., Pirulli, M., & Scavia, C. (2010). Microseismic activity analysis for the study of the rupture mechanism in unstable rock masses. *Natural Hazards and Earth System Science, 10*, 831–841.

Arosio, D., Longoni, L., Papini, M., Scaioni, M., Zanzi, L., & Alba, M. (2009a). Towards rockfall forecasting through observing deformations and listening to microseismic emissions. *Natural Hazards and Earth System Science, 9*, 1119–1131.

Arosio, D., Bernasconi, G., Mazzucchelli, P., Rovetta, D., & Zanzi L. (2009b). Localization algorithms for search and rescue applications. In *Exp. Abs. Near Surface 2009*, Dublin, Ireland, September 7–9, 2009.

Arosio, D. (2010). A microseismic approach to locate survivors trapped under rubble. *Near Surface Geophysics, 8*, 623–633.

Arosio, D., Longoni, L., Mazza, F., Papini, M., & Zanzi, L. (2013). Freeze-thaw cycle and rockfall monitoring. In C. Margottini, et al. (Eds.), *Landslide science and practice* (Vol. 2, pp. 385–390). Berlin Heidelberg: Springer.

Arosio, D., Longoni, L., Tarabini, M., Papini, M., Zanzi, L., & Colombo, M. (2014). Microseismic monitoring of rockfalls: Preliminary interpretation of lab and field tests. In *Proceedings of 5th Interdisciplinary Workshop on Rockfall Protection (RocExs 2014)*. Lecco, Italy, May 29–31, 2014.

Blikra, L. H. (2008). The Åknes rockslide: Monitoring, threshold values and early warning. In *Proceedings of 10th International Symposium on Landslides and Engineered Slopes* (pp. 1089–1094). Xi'an, P.R. China, June 30–July 4, 2008.

Blikra, L. H. (2012). The Åknes rockslide, Norway. In J. J. Clague & D. Sead (Eds.), *Landslides: Types mechanisms and modeling* (pp. 323–335). UK: Cambridge University Press.

Drew, J., Leslie, D., Armstrong, P., & Michaud, G. (2005). In *Proceedings of SPE Annual Technical Conference and Exhibition* (paper no. SPE95513). Dallas, TX, USA, October 9–12, 2005.

Hardy, H. R. (2003). *Acoustic emission/microseismic activity—Principles, techniques and geotechnical applications*. Lisse, The Netherlands: A.A. Balkema Publishers.

Helmstetter, A., & Garambois, S. (2010). Seismic monitoring of Séchilienne Rockslide (French Alps): Analysis of seismic signals and their correlation with rainfalls. *Journal of Geophysical Research 115*, Paper no. F03016.

Herath, S. C. K., & Pathirana, P. N. (2013). Robust localization with minimum number of TDOA measurements. *IEEE Signal Processing Letters, 20*(10), 949–951.

Senfaute, G., Duperret, A., & Lawrence, J. A. (2009). Micro-seismic precursory cracks prior to rock-fall on coastal chalk cliffs: A case study at Mesnil-Val, Normandie, NW France. *Natural Hazards and Earth System Science, 9*, 1625–1641.

Spillmann, T., Maurer, H., Green, A. G., Heincke, B., Willenberg, H., & Husen, S. (2007). Microseismic investigation of an unstable mountain slope in the Swiss Alps. *Journal of Geophysical Research 112*, Paper no. B07301.

Walter, M., Schwaderer, U., & Joswig, M. (2012). Seismic monitoring of precursory fracture signals from a destructive rockfall in the Vorarlberg Alps, Austria. *Natural Hazards and Earth System Science, 12*, 3545–3555.

The State of the Art of SPH Modelling for Flow-slide Propagation

Zili Dai and Yu Huang

Abstract Flow-slide disaster is a continuing problem along hillsides in mountainous areas, which always results in numerous casualties and catastrophic destruction of buildings and regional landscapes. Predicting of the propagation stage is of great importance for the disaster mitigation. The smoothed particle hydrodynamics (SPH) method, a mesh-free particle technique, has been widely applied for modelling of flow-slide evolution with some success. The main goal of this chapter was to provide a general view of SPH applications for the analysis of flow-slide disasters including flow-like landslides, landslide-generated waves, and debris flows. The leading features of the proposed SPH models are detailed and the achievements are presented and discussed.

Keywords Flow-slide · Numerical modelling · Smoothed particle hydrodynamics · Propagation stage

1 Introduction

Flow-slide is a serious frequently occurred geological disaster. It includes flow-like landslides, debris flows, rock avalanches, and so on. Once occurred, these disasters often result in catastrophic events due to the suddenness and rapid evolution.

Z. Dai · Y. Huang (✉)
Department of Geotechnical Engineering, College of Civil Engineering, Tongji University, Shanghai 200092, People's Republic of China
e-mail: yhuang@tongji.edu.cn

Z. Dai
e-mail: 1040020033@tongji.edu.cn

Y. Huang
Key Laboratory of Geotechnical and Underground Engineering of the Ministry of Education, Tongji University, Shanghai 200092, People's Republic of China

M. Scaioni (ed.), *Modern Technologies for Landslide Monitoring and Prediction*, Springer Natural Hazards, DOI 10.1007/978-3-662-45931-7_8

Therefore, they generate a yearly loss of property larger than the one coming from other natural disasters.

Prediction of the runout distance and velocity through numerical modelling of the propagation stage may significantly reduce losses caused by such disasters, as it can provide a means to define the hazardous area and estimate the disaster intensity. In order to conduct a quantitative risk analysis for flow slides, different numerical models with various accuracies have been developed over the past decades. Most of them were based on grids, such as finite element method (FEM), finite difference method (FDM), and finite volume method (FVM). All these approaches were effective in solving partial differential equations (PDE) and obtained interesting results in several studies of flow-slide disasters. For example, Crosta et al. (2003, 2004, 2006) used an FEM model to simulate flow-like landslides. In fact, numerical techniques based on fixed or moving meshes are poorly suited for simulating the propagation of flow slides, which are always accompanied by extremely large deformations, free surfaces, and moving interfaces.

An interesting and robust alternative to these methods is provided by a new group of 'mesh-free' techniques, which were developed and improved in the past decades. Among these methods, the *discrete element method* (Cundall and Strack 1979), the *diffuse element method* (Nayroles et al. 1992), the *element-free Galerkin method* (Belytschko et al. 1994), the *reproducing kernel particle method* (Liu et al. 1995), the moving particle semi-implicit method (Koshizuka and Oka 1996), and finally the smoothed particle hydrodynamics (SPH) method (Lucy 1977; Gingold and Monaghan 1977), which is the technique described in this chapter, are worth mentioning.

The SPH is a recently developed 'mesh-free' technique. Due to the natural feature of 'mesh-free' and Lagrangian description, this method has been successfully applied in a range of fields; exhaustive reviews can be found in Monaghan (1992) and Liu and Liu (2010). The objective of this paper is to detail some features associated with SPH simulations for propagation stage of flow-slide disasters. An overview of SPH applications to flow-like landslides (Sect. 2.1), landslide-generated waves (Sect. 2.2), and debris flows (Sect. 2.3) is presented. The achievements in the related literatures are discussed and conclusions are drawn in Sect. 3.

2 SPH Applications to Flow Slides

In recent years, many SPH models have been proposed for the modelling and analysis of flow-slide disasters. In this section, the SPH applications to the flow-like landslide, landslide-generated wave and debris flow are reviewed. The key features of the presented SPH models are detailed and the achievements are presented and discussed.

2.1 Flow-like Landslides

Under the influence of *natural factors* (earthquakes, rainfall, and erosion) and *anthropogenic factors* (excavation, road-cuts, construction and reshaping works, etc.), landslides may occur in natural slopes and man-made structures including embankment dams, hydraulic fills, waste landfills, tailing dams, and coal dumps.

Once occurred, landslides may propagate with velocities ranging from few centimeters per month to the order of a hundred kilometers per hour. The propagation velocity depends largely on the rheological properties of the involved materials and its failure mechanism. In certain cases, the material moves quite fast in large volumes, reaches a long travel distance and behaves in a fluid manner, and it is always associated with serious damage and huge casualties. This kind of landslide is defined as *flow-like landslide* (Hungr et al. 2001). Understanding the failure mechanism and predicting both runout distances and velocity through mathematical modelling of the propagation stage is important to provide a means for defining the hazardous areas as well as for estimating the hazard intensity.

To the authors' knowledge, the first work that described the SPH application to flow-like landslides modelling was presented by McDougall and Hungr (2004). A numerical model was developed to capture the major characteristics of rapid landslide motion across 3D terrain. The depth-integrated equations of shallow-fluid flow were introduced as governing equations to describe a column of earth material moving with the landslide and discretized by the SPH method. The heterogeneous and complex landslide material was assumed to be a viscous material, and its resistance parameters were selected based on trial-and-error back-analysis of full-scale landslides. This model was applied to model the evolution of the Frank Slide (Alberta, Canada), and promising results were obtained. Then, a *path material entrainment* algorithm was incorporated in the model (McDougall and Hungr 2005), and this capability was demonstrated using a back-analysis of the 1999 Nomash River landslide (British Columbia, Canada). The unique capabilities of the proposed model include the ability to account for material entrainment, rheology variation, and complex internal stress states. However, the applicability to the deep-seated landslides with highly irregular paths should be further investigated.

Since then, Pastor et al. (2009) proposed a depth-integrated, coupled SPH model to simulate the propagation of flow-like catastrophic landslides. The main difference from the McDougall and Hungr's model was that a Biot–Zienkiewicz model in the velocity–pressure version was introduced to consider the coupling effect between the solid and fluid phases. As an application, the propagation stage of the flow-like landslide (May 1998) located in southern Italy was simulated, and the output results were consistent with the available field observations. After that, Haddad et al. (2010) applied this model to simulate the propagation stage of a *lahar* in Mexico. Meanwhile, parameter sensitivity analysis of the proposed model was conducted, which showed that the yield strength had a strong influence on the runout distance while the viscosity affected mainly the flow velocity. The main advantage of this model is the ability to take into account pore-water-pressure dissipation during propagation, which plays an important role in the strength recovery of the materials. More recently, the same group (Pastor et al. 2014) implemented rheological models to describe basal friction and a simple erosion law to consider the bed entrainment phenomenon, and showed the potential of depth-integrated, SPH models to simulate the propagation phase of different types of landslides. Cascini et al. (2014) applied the depth-integrated-coupled SPH model to simulate the propagation stage of rainfall-induced channelized landslides, and clearly indicated that the bed entrainment rate and the extent of the erodible areas affect the flow paths, the velocities, and the final deposition heights.

On the other hand, Bui et al. (2008a) introduced an elastic–plastic constitutive model of soil called Drucker–Prager model to the SPH framework for the failure flows simulation of geo-materials. An artificial stress method first proposed by Monaghan (2000) was introduced to remove the SPH numerical instability. In addition, they presented an algorithm to deal with the problem of soil-structure interaction, see Bui et al. (2008b). The key peculiarities of this research are the implementation of an elastic–plastic soil constitutive model into an SPH model, and the implementation of a well-performing contact algorithm to deal with soil-structure interactions.

Our group presented an SPH model for the runout prediction of flow-like landslides, see Huang et al. (2012). Different from the depth-integrated, this model applied Navier–Stokes equations as the governing equations to describe the soil-flow characteristics. The landslide material was assumed as a kind of Bingham fluid, and *Mohr-Coulomb yield criterion* was incorporated into the Bingham model to calculate the soil yield strength. Visual software with user-friendly interfaces was successfully developed, improving the simulation efficiency, see Huang et al. (2011). This model simulated landslide motion along a user-prescribed two-dimensional (2D) path, which may have contributed to the error budget. Recently, this model was extended to the three-dimensional (3D) domain, and the no-slip boundary condition was incorporated to simulate the 3D complex terrain (Dai and Huang 2014). In addition, a parallel algorithm based on Open Multiprocessing (OpenMP) was used to improve the efficiency of the SPH code running. The propagation stages of typical flow-like landslides induced by the 12th May, 2008, Wenchuan earthquake were simulated, and the runout and deposition area predicted had a high degree of similarity with real landslide configurations as surveyed during field inspection. Figure 1 shows the simulated propagation of the Tangjiashan landslide, and the corresponding calculating parameters are provided in Table 1. The main contribution of this work was the incorporation of the *soil strength criterion* in the SPH model and the introduction of the concept of *equivalent viscosity* to improve the Bingham model. It demonstrated that the soil in flow condition can be described as a viscous fluid.

Unlike traditional numerical models based on solid mechanics, the SPH models mentioned above analyze the dynamic behavior of slopes from a fluid-mechanics point of view and provide a completely new and effective approach for runout prediction. In the SPH models above, complex constitutive laws of geo-materials have been used, especially the viscoplastic fluid named *Bingham fluid* (Ataie-Ashtiani and Shobeyri 2008; Pastor et al. 2009; Haddad et al. 2010; Capone et al. 2010;

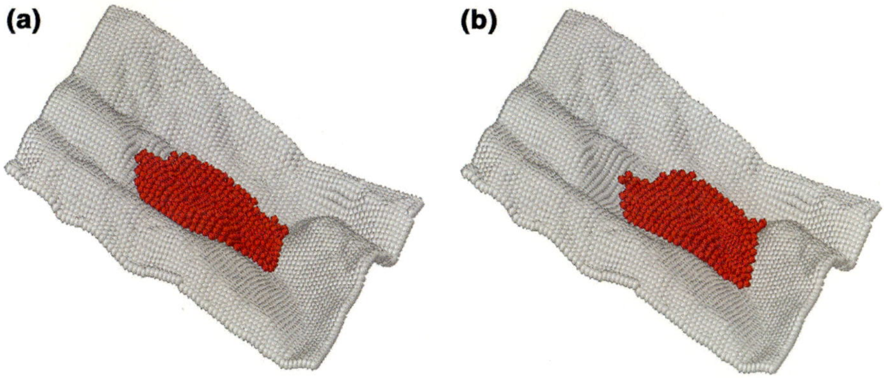

Fig. 1 Simulated propagation of the Tangjiashan landslide: on the left (**a**) after t = 10 s; on the right (**b**) after t = 20 s

Table 1 Parameters used in the propagation simulation of the Tangjiashan landslide

Density	ρ (kg/m^3)	2,000
Cohesion	c (kPa)	30
Angle of internal friction	φ (°)	30
Viscosity coefficient	η (Pa s)	1.9
Acceleration of gravity	g (m/s^2)	9.8
Unit time step	Δt (s)	2.0×10^{-4}
Time step	N	2.0×10^5

Huang et al. 2011, 2012). A simpler semi-empirical approach based on the concept of 'equivalent fluid' was used in a new SPH model by McDougall and Hungr (2004) in a landslide study. The incorporation of these complex constitutive models in the SPH framework promotes the application of this method in geo-disasters analysis.

2.2 *Landslide-generated Waves*

Landslide-generated waves (also referred to as 'landslide tsunamis') occur in the ocean or restricted water bodies, such as lakes, reservoirs, or fjords, have a high potential to cause damage and loss of life. Understanding the generation mechanism and predicting the propagation stage of these waves are important for assessing the level of potential flooding.

Schwaiger and Higman (2007) tested the feasibility of using an SPH model to simulate the tsunami generated by the 1958 Lituya Bay rockslide (Alaska, USA). In this model, rock and water were treated as an incompressible Newtonian viscous and inviscid fluid, respectively. Pressure of the fluids was determined through an equation of state. The simulation results from this model resulted in good agreement with field observations and a scaled-down laboratory experiment. A 2D SPH model for inviscid fluid was developed by Qiu (2008) to simulate landslide-induced waves, and the generation and propagation of the water-triggered triangular rigid body was accurately predicted. The comparison with experimental data demonstrated that SPH is an appropriate model to simulate a landslide-generated water wave. SPH modelling of tsunami-wave generation was carried out by Das et al. (2009); the complex flow patterns predicted in terms of the free surface deformation, shoreline movement, and wave run-up shoreline evolution were fully compatible with experimental data and the numerical results from the Reynolds averaged Navier–Stokes (RANS) model. Capone et al. (2010) introduced a numerical representation of the rheological non-Newtonian Bingham model by means of the SPH method to investigate the landslide deformation and its interaction with water, thus reproducing the generation and propagation of a tsunami triggered by underwater landslides.

Due to the easy implementation of the conservation laws of motion, some Eulerian techniques, such as the volume-of-fluid (VOF) method and the level-set (LS) approach, have been incorporated into SPH models to track moving interfaces between different fluids. For example, Abadie et al. (2010) incorporated the VOF algorithm into a 3D multiple-fluid SPH model for the simulation of landslide-generated waves. The wave, air, and slope were treated as different Newtonian fluids and the interactions were tracked using the VOF technique. This model was validated by comparison with semi-empirical laws of motion and experimental data, and successfully applied in simulations of retrogressive slope failure, deformable slides, and granular flows. Another widely used interface-capturing method is the LS approach. For example, Marrone et al. (2010) defined a LS function in the SPH solver to detect the fast free surface.

The conventional SPH method was extended to incompressible or nearly incompressible flow by using two different approaches to enhance numerical stability and accuracy. The first approach is weakly compressible smoothed particle hydrodynamics (WCSPH) method, which uses equations of state to calculate the pressure, see Monaghan (1994). This method was corrected and applied by Ataie-Ashtiani and Mansour-Rezaei (2009) to simulate an impulsive wave generated by an underwater landslide. Though this method is easy to be implemented into numerical code, it has some drawbacks, one of which is that sound wave reflection will appear near the boundary caused by the artificial compressibility, see Shao and Lo (2003). This problem can be overcome by *incompressible-SPH* (I-SPH) method. Here, pressure is computed from a Poisson equation which satisfies the incompressibility condition. It is easy and efficient to track the free surface and treat wall boundaries. For example, Ataie-Ashtiani and Shobeyri (2008) developed an I-SPH model to simulate a submerged rigid wedge sliding along an inclined surface, and the results were in good agreement with the experimental data. Comparisons of the I-SPH algorithm with the classical WCSPH method were presented by Lee et al. (2008), who showed that I-SPH yields much more reliable results than WCSPH.

2.3 Debris Flows

Similar to landslides, debris flows and mudflows belong to rapid gravity-driven flow slides and usually burst out suddenly and unpredictably. They are unsteady flows of highly concentrated mixtures of water and solid material with large grain size distribution, which can always cause serious damage to human infrastructures.

Rodriguez-Paz and Bonet (2004) presented a numerical model for debris flow based on a corrected smooth particle hydrodynamics (CSPH) code. This model introduced an adjusting factor and a correction matrix to correct the kernel functions and their gradients. The debris flow was represented as a single material, and the Bingham model was used as constitutive law. A new frictional approach for the boundary conditions was introduced in the SPH method. The resulting model was applied to simulate a debris-flow experiment, and satisfactory results were achieved.

Minatti and Pasculli (2011) developed a 2D muddy debris-flow model based on SPH method. In this model, the solid–liquid two-phase flow was considered as one equivalent material through the *debris-flow mixture theory*. Then the Herschel-Bulkley fluid model was selected as constitutive law to correctly describe the behavior of mudflows. The ability of the SPH model to correctly reproduce a mudflow was demonstrated by simulating mudflow dam break problem in a laboratory flume. More recently, this group improved the SPH model for fast muddy debris flow, see Pasculli et al. (2013). To avoid unphysical high-frequency density oscillations, a flux term devised from the Rusanov flux was added to the continuity equation, and a Morris viscous term was incorporated in the momentum equation for the simulation of incompressible viscous flows. The application of such new model in more than thirty numerical experiments provided satisfactory results.

Fang et al. (2013) designed a fluid-solid coupling SPH model, in which the debris flow was simplified as fluid and rigid-body particles. The impact and vortex force exerted on the rigid body by the fluid particles could be taken into consideration. The applicability of this comprehensive method to simulate realistic debris flows was proved. The highlight of this research is that the proposed SPH model can consider the effect of fluid-solid coupling during the debris-flow propagation, and the rotation algorithm was used to optimize the movement simulation of the rigid body.

Debris flows and mudflows can damage human infrastructures due to the impact force exerted by the flow front. Many structures protecting against debris flows and mudflows are installed in torrent channels. Therefore, it is of paramount importance to understand debris-flow structure interactions for the rational design of these retaining structures. To this topic, Laigle et al. (2007) proposed a numerical approach based on SPH method. An equation of state was used to calculate the pressure value on the basis of density. Herschel-Bulkley fluid model was introduced as the material law. They defined the instantaneous momentum, and the momentum reduction due to structure was calculated. Appling this model, some debris-flow model tests were simulated. Though the results presented are valuable, the model still needs to be validated by further numerical investigations.

3 Conclusions

Flow slides (e.g., flow-like landslides, debris flows, and rock avalanches) could easily lead to casualties and large-scale economic losses. It is important to predict propagation stage of the disasters and then to define the hazard area. This paper reviewed numerical models based on the SPH method for the studies of flow-like landslides, landslide-generated waves, and debris flows. From the achievements of the studies reviewed in this paper, the SPH method has significant advantages over traditional grid-based numerical modelling methods when dealing with flow-slide disasters concerning extremely large deformation, free surfaces, and moving interfaces. This demonstrates that SPH is a promising method for studying such kind of geo-disasters.

Acknowledgment This work was supported by the National Basic Research Program of China (973 Program) through Grant No. 2012CB719803.

References

Abadie, S., Morichon, D., Grilli, S., & Glockner, S. (2010). Numerical simulation of waves generated by landslides using a multiple-fluid Navier-Stokes model. *Coastal Engineering, 57*, 779–794.

Ataie-Ashtiani, B., & Mansour-Rezaei, S. (2009). Modification of weakly compressible smoothed particle hydrodynamics for preservation of angular momentum in simulation of impulsive wave problems. *Coastal Engineering Journal, 51*(4), 363–386.

Ataie-Ashtiani, B., & Shobeyri, G. (2008). Numerical simulation of landslide impulsive waves by incompressible smoothed particle hydrodynamics. *International Journal for Numerical Method, 56*, 209–232.

Belytschko, T., Lu, Y. Y., & Gu, L. (1994). Element-free Galerkin methods. *International Journal for Numerical Methods in Engineering, 37*, 229–256.

Bui, H. H., Fukagawa, R., Sako, K., & Ohno, S. (2008a). Lagrangian meshfree particles method (SPH) for large deformation and failure flows of geomaterial using elastic–plastic soil constitutive model. *International Journal for Numerical and Analytical Methods in Geomechanics, 32*, 1537–1570.

Bui, H. H., Sako, K., Fukagawa, R., & Wells, J. C. (2008b). SPH-based numerical simulations for large deformation of geomaterial considering soil-structure interaction. In *Proceedings 12th International Conference of International Association for Computer Methods and Advances in Geomechanics* (pp. 1–6), Goa, India, October 1–6, 2008.

Capone, T., Panizzo, A., & Monaghan, J. J. (2010). SPH modeling of water waves generated by submarine landslides. *Journal of Hydraulic Research, 48*, 80–84.

Cascini, L., Cuomo, S., Pastor, M., Sorbino, G., & Piciullo, L. (2014). SPH run-out modelling of channelised landslides of the flow type. *Geomorphology, 214*, 502–513.

Crosta, G. B., Imposimato, S., & Roddeman, D. G. (2003). Numerical modelling of large landslides stability and runout. *Natural Hazards and Earth System Sciences, 3*, 523–538.

Crosta, G. B., Imposimato, S., Roddeman, D. G., Chiesa, S., & Moia, F. (2004). Small fast moving flow-like landslides in volcanic deposits: The 2001 Las Colinas Landslide (El Salvador). *Engineering Geology, 79*, 185–214.

Crosta, G. B., Imposimato, S., & Roddeman, D. G. (2006). Continuum numerical modelling of flow-like landslides. *Landslides from Massive Rock Slope Failure, 49*, 211–232.

Cundall, P. A., & Strack, O. D. L. (1979). A discrete numerical model for granular assemblies. *Geotechnique, 29*, 47–65.

Dai, Z. L., & Huang, Y. (2014). 3D numerical modeling using smoothed particle hydrodynamics of flow-like landslide propagation triggered by the 2008 Wenchuan earthquake. *Engineering Geology*. doi:10.1016/j.enggeo.2014.03.018.

Das, K., Janetzke, R., Basu, D., Green, S., & Stamatakos, J. (2009). Numerical simulations of tsunami wave generation by submarine and aerial landslides using RANS and SPH models. In *Proceedings 28th International Conference on 'Ocean, Offshore and Arctic Engineering'* (Vol. 5, pp. 581–594), Honolulu, HI, USA.

Fang, H., Sun, S., Wang, X., & Chen, D. (2013). A simulation animation method of debris flow based on improved SPH and fixed-point rotation. *International Journal of Digital Content Technology and its Applications, 7*(2), 721–730.

Gingold, R. A., & Monaghan, J. J. (1977). Smoothed particles hydrodynamics: theory and application to non-spherical stars. *Monthly Notices of the Royal Astronomical Society, 181*, 375–389.

Haddad, B., Pastor, M., Palacios, D., & Munoz-Salinas, E. (2010). A SPH depth integrated model for Popocatepetl 2001 lahar (Mexico): Sensitivity analysis and runout simulation. *Engineering Geology, 114*(3–4), 312–329.

Huang, Y., Dai, Z. L., Zhang, W. J., & Chen, Z. Y. (2011). Visual simulation of landslide fluidized movement based on smoothed particle hydrodynamics. *Natural Hazards, 59*(3), 1225–1238.

Huang, Y., Zhang, W. J., Xu, Q., Xie, P., & Hao, L. (2012). Run-out analysis of flow-like landslides triggered by the Ms 8.0 2008 Wenchuan earthquake using smoothed particle hydrodynamics. *Landslides, 9*(2), 275–283.

Hungr, O., Evans, S. G., Bovis, M. J., & Hutchinson, J. N. (2001). A review of the classification of landslides of the flow type. *Environmental and Engineering Geoscience, 7*(3), 221–238.

Koshizuka, S., & Oka, Y. (1996). Moving-particle semi-implicit method for fragmentation of incompressible fluid. *Nuclear Science and Engineering, 123*(3), 421–434.

Laigle, D., Lachamp, P., & Naaim, M. (2007). SPH-based numerical investigation of mudflow and other complex fluid flow interactions with structures. *Computational Geosciences, 11*(4), 297–306.

Lee, E. S., Moulinec, C., Xu, R., Violeau, D., Laurence, D., & Stansby, P. (2008). Comparisons of weakly compressible and truly incompressible algorithms for the SPH mesh free particle method. *Journal of Computational Physics, 227*, 8417–8436.

Liu, W. K., Jun, S., & Zhang, Y. F. (1995). Reproducing kernel particle methods. *International Journal for Numerical Methods in Fluids, 20*(8–9), 1081–1106.

Liu, M. B., & Liu, G. R. (2010). Smoothed particle hydrodynamics (SPH): An overview and recent developments. *Archives of Computational Methods in Engineering, 17*(1), 25–76.

Lucy, L. B. (1977). A numerical approach to the testing of fusion process. *Astronomical Journal, 82*, 1013–1024.

Marrone, S., Colagrossi, A., Le Touze, D., & Graziani, G. (2010). Fast free-surface detection and level-set function definition in SPH solvers. *Journal of Computational Physics, 229*(10), 3652–3663.

McDougall, S., & Hungr, O. (2004). A model for the analysis of rapid landslide motion across three-dimensional terrain. *Canadian Geotechnical Journal, 41*(6), 1084–1097.

McDougall, S., & Hungr, O. (2005). Dynamic modelling of entrainment in rapid landslides. *Canadian Geotechnical Journal, 42*(5), 1437–1448.

Minatti, L., & Pasculli, A. (2011). SPH numerical approach in modelling 2D muddy debris flow. In *Proceedings of 5th International Conference on 'Debris-Flow Hazards Mitigation: Mechanics, Prediction, and Assessment* (*5th DFHM*)' (pp. 467–475), Padua, Italy, June 14–17, 2011.

Monaghan, J. J. (1992). Smoothed particle hydrodynamics. *Annual Reviews of Astronomy and Astrophysics, 30*, 543–574.

Monaghan, J. J. (1994). Simulating free-surface flows with SPH. *Journal of Computational, 110*(2), 399–406.

Monaghan, J. J. (2000). SPH without a tensile instability. *Journal of Computational Physics, 159*(2), 290–311.
Nayroles, B., Touzot, G., & Villon, P. (1992). Generalizing the finite element method: Diffuse approximation and diffuse elements. *Computational Mechanics, 10*, 307–318.
Pasculli, A., Minatti, L., Sciarra, N., & Paris, E. (2013). SPH modeling of fast muddy debris flow: Numerical and experimental comparison of certain commonly utilized approaches. *Italian Journal of Geosciences, 132*(3), 350–365.
Pastor, M., Blanc, T., Haddad, B., Petrone, S., Sanchez Morles, M., Drempetic, V., et al. (2014). Application of a SPH depth-integrated model to landslide run-out analysis. *Landslides*. doi:10.1007/s10346-014-0484-y.
Pastor, M., Haddad, B., Sorbino, G., & Drempetic, V. (2009). A depth-integrated, coupled SPH model for flow-like landslides and related phenomena. *International Journal for Numerical and Analytical Methods in Geomechanics, 33*, 143–172.
Qiu, L. C. (2008). Two-dimensional SPH simulations of landslide-generated water waves. *Journal of Hydraulic Engineering, 134*(5), 668–671.
Rodriguez-Paz, M. X., & Bonet, J. (2004). A corrected smooth particle hydrodynamics method for the simulation of debris flows. *Numerical Methods for Partial Differential Equations, 20*(1), 140–163.
Schwaiger, H.F., & Higman, B. (2007). Lagrangian hydrocode simulations of the 1958 Lituya Bay tsunamigenic rockslide. *Geochemistry Geophysics Geosystems, 8*(7), paper no. Q07006.
Shao, S. D., & Lo, E. Y. M. (2003). Incompressible SPH method for simulating Newtonian and non-Newtonian flows with a free surface. *Advances in Water Resources, 26*(7), 787–800.

Predictability of a Physically Based Model for Rainfall-induced Shallow Landslides: Model Development and Case Studies

Yang Hong, Xiaogang He, Amy Cerato, Ke Zhang, Zhen Hong and Zonghu Liao

Abstract A cost-effective physical model (SLope-Infiltration-distributed Equilibrium—SLIDE) has been developed to identify the spatial and temporal occurrences of rainfall-induced landslides, employing a range of remotely sensed and in situ data. The main feature of SLIDE is that it takes into account of some simplified hypotheses on water infiltration and defines a direct relationship between the factor of safety and the rainfall depth on an infinite slope model. This prototype has been applied to two case studies in Indonesia and Honduras during heavy rainfall events brought by typhoon and hurricane, respectively. Simulation results from SLIDE demonstrated good skills in predicting rainfall-induced shallow landslides by assimilating the most important dynamic triggering factor (i.e., rainfall) quantitatively. The model's prediction performance also suggested that SLIDE could serve as a potential tool for the future landslide early-warning system. Despite positive model performance, the SLIDE model is limited by several

Y. Hong · X. He · A. Cerato · Z. Hong · Z. Liao
Department of Civil Engineering and Environmental Science,
and Advanced Radar Research Center, The University of Oklahoma,
Norman, OK 73072, USA
e-mail: hexg@princeton.edu

A. Cerato
e-mail: acerato@ou.edu

Z. Hong
e-mail: Zhen.Hong-1@ou.edu

Z. Liao
e-mail: zonghuliao@ou.edu

Y. Hong (✉)
National Weather Center ARRC Suite 4610, 120 David L. Boren Blvd.,
Norman, OK 73072, USA
e-mail: yanghong@ou.edu

K. Zhang
Cooperative Institute for Mesoscale Meteorological Studies,
The University of Oklahoma, Norman, OK, USA
e-mail: kezhang@ou.edu

M. Scaioni (ed.), *Modern Technologies for Landslide Monitoring and Prediction*,
Springer Natural Hazards, DOI 10.1007/978-3-662-45931-7_9

assumptions including using general parameter calibration rather than in situ tests and neglecting geotechnical information and some of the hydrological processes in deep soil layers. Advantages and limitations of this physically based model are also discussed with respect to future applications of landslide assessment and prediction over large scales.

Keywords Physically based models · Prediction · Rainfall-induced · Shallow landslides · SLIDE · TRMM

1 Introduction

Rainfall-induced landslides are one of the most important disasters to occur in complex terrain areas, especially regions that routinely experience heavy rainfall. Some of these landslides occur suddenly and travel at high speeds, posing significant threats to life and property (Iverson 2000; Hong et al. 2006; Kirschbaum et al. 2009a). Overpopulation, deforestation, mining, and uncontrolled land-use for agricultural and transportation purposes increasingly put large numbers of people at risk from landslides (Boebel et al. 2006; Sidle and Ochiai 2006). It is important to approach landslide susceptibility, hazard, and risk assessment at the appropriate spatial scales to inform policy decisions and proper planning for mitigating risk. Landslide disaster preparedness and forecasting systems require knowledge of the physical mechanisms influencing slope failure, information on geotechnical composition of soil, and landslide inventory data.

Recent approaches to forecast landslide potential include *empirical models* that are based on rainfall intensities and duration required to trigger landslides (i.e., Dai et al. 2002; Hong et al. 2007). These empirical models provide little theoretical basis for understanding how landslides might respond to hydrologic processes (Iverson 2000). For most shallow landslides, rainfall triggers slope failure because water reduces the shear strength and increases the shear stress in the soil layer. Compared to empirical models, *physically based landslide models* consider the physical mechanisms influencing slope instability to assess landslide hazard, using a range of topographic, geologic, geotechnic, and hydrologic parameters (Dietrich et al. 1995; Wu and Sidle 1995; Baum et al. 2002; Iverson 2000; Lu and Godt 2008; Godt et al. 2009). These physical models generally employ high-resolution surface feature data, in situ geotechnical information, rainfall measurements at the land surface, and detailed landslide inventories. Such information is frequently unavailable at larger spatial scales due to non-uniform surface investigation and sparse landslide inventory data (Kirschbaum et al. 2012).

Recent examinations of remote-sensing data sets offer an opportunity to enhance the sparse field-based landslide inventory data sets for hazard forecasting and mapping at larger spatial scales. In this article, a *simplified physical model*, SLope-Infiltration-Distributed Equilibrium (SLIDE) has been developed to identify the spatial and temporal distribution of landslides induced by heavy rainfall, employing a range of remotely sensed and in situ surface data. Utilizing satellite remote-sensing data in landslide studies enables researchers to develop an operational early warning system not only at the site scale, but also at the regional scale.

The remainder of this article is organized as follows: Sect. 2 introduces the framework of the SLIDE model including its model structures, data collection, and parameter initialization. In Sect. 3, application and validation of this model at regional scale in Indonesia (Liao et al. 2010) and Honduras (Liao et al. 2011) are described as case studies to test its potential for regional landslide risk assessment. Finally, limitations of the model and its potential improvement are briefly discussed at the end of this chapter.

2 Methodology

2.1 Model Development

The SLIDE model takes into account some simplified hypotheses on water infiltration and defines a direct relationship between the factor of safety (FS) and the rainfall depth on an infinite slope model (Taylor 1948) (Fig. 1). This prototype model can provide landslide mapping and forecasting through the utilization of remotely sensed and in situ surface data at larger scales.

The SLIDE model integrates the contribution of apparent cohesion to the shear strength of the soil and soil depth influenced by water infiltration processes. In Fig. 1, *N* is normal effective force, *T* is the shear force, and t_i is the time step of infiltration. FS is expressed as the ratio of shear strength to shear stress to calculate slope stability. A link between the rainfall amount and the final expression of FS has been set up and translated into a simple mathematical formulation as shown in the following Eq. (1). A slope is considered stable when FS > 1 and a landslide are predicted to occur when FS nears or drops below this threshold. SLIDE assumes

Fig. 1 Schematic diagram illustrating the SLIDE model of the infinite slope

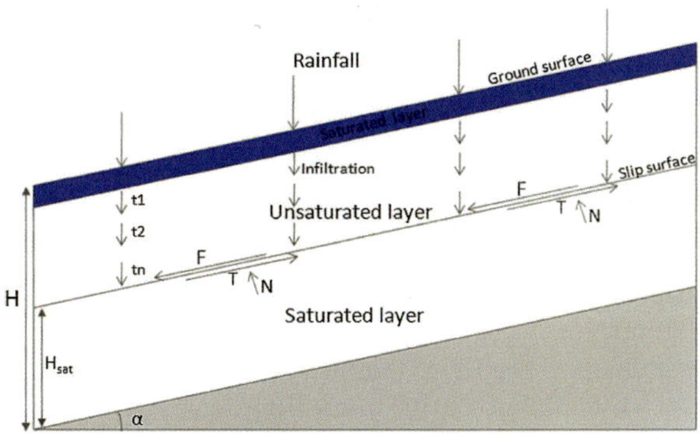

that landslides occur in shallow depth and an infinite slope equation is translated as the cohesion and frictional components:

$$\mathrm{FS}(Z_t, t) = \frac{c' + c_\phi(t)}{\gamma_s Z_t \sin\alpha\cos\alpha} + \frac{\tan\varphi}{\tan\alpha} \tag{1}$$

where c' is soil cohesion, incorporating a value for root zone cohesion, γ_s is the unit weight of soil, α is slope angle, and φ is soil friction angle. $c_\varphi(t)$ represents the apparent cohesion related to matric suction, which in turn depends on the degree of saturation of the soil (Montrasio and Valentino 2008), written as follows:

$$c_\phi(t) = A \cdot S_r \cdot (1 - S_r)^\lambda \cdot (1 - m_t)^\delta \tag{2}$$

where A is a parameter depending on the soil type that is linked to the peak shear stress at failure, λ and δ are numerical parameters which allow estimation of the peak of apparent cohesion related to S_r, the degree of soil saturation, m_t represents the dimensionless thickness of the infiltrated layer, which is a fractional parameter between 0 and 1:

$$m_t = \frac{\sum_{t=1}^{T} I_t}{n \cdot Z_t \cdot (1 - S_r)} \tag{3}$$

in which I_t is rain intensity, n is the porosity, and Z_t is the soil depth at time t, which is determined by the infiltration processes:

$$Z_t = \sqrt{\frac{2 \cdot K_s \cdot H_c \cdot t}{\theta_n - \theta_0}} \tag{4}$$

where K_s is saturated hydraulic conductivity, H_c is capillary pressure, θ_n is water content of the saturated soil, and θ_0 is initial water content of the soil.

2.2 Data Collection

2.2.1 Land Surface Data and Parameters

Landslide occurrence depends on complex interactions among a large number of factors, mainly including slope, soil properties, lithology, and land cover. With increasing frequency, remote-sensing data sets have been used to develop susceptibility maps at regional (He et al. 2014) and global scales (Hong et al. 2007). In this chapter, elevation, land cover, and precipitation data sets were all derived from satellites as inputs of the SLIDE model. Topographic data were retrieved from the ASTER Global Digital Elevation Model (DEM), which was developed jointly by

Table 1 Parameters and symbols used in SLIDE model

Property	Symbol	Unit
Slope angle	α	Degree
Soil depth	Z_t	m [L]
Soil type	/	Unitless
Land cover	/	Unitless
Coefficients	λ, δ	Unitless
Friction	φ	Degree
Cohesion (root included)	c'	kPa (M/LT^2)
Coefficient	A	kPa (M/LT^2)
Unit weight of soil	γ_s	KN/m^3 (M/L^2T^2)
Porosity	n	m^3/m^3
Water content	θ	m^3/m^3
Degree of saturation	S_r	Fraction
Hydrologic conductivity	K_s	m/s (L/T)
Capillary	H_c	m (L)

the Ministry of Economy, Trade, and Industry of Japan and the United States NASA (https://wist.echo.nasa.gov/~wist/api/imswelcome/). Topographic properties were derived from a 30 m × 30 m ASTER DEM. Soil parameter values were determined by referring to the geotechnical literature according to soil types provided by the Food and Agriculture Organization of the United Nations (FAO; http://www.fao.org/AG/agl/agll/dsmw.htm) and the Moderate Resolution Imaging Spectroradiometer (MODIS) land classification map.

The application of the SLIDE model requires the assignment of 14 parameters, as summarized in Table 1. Since such parameters have a large influence on SLIDE's performance, it is imperative to assign values as realistic as possible based on field surveys and laboratory tests. However, due to the limited soil data from field measurements, values and coefficients of soil have been derived through a procedure of back analysis within a certain range and validated by inventory data. Regarding conceptual parameters (A, λ, and δ), a parameter-calibration approach is proposed to calibrate the model based on the reported landslide inventory database.

2.2.2 Precipitation

The spatial distribution, duration, and intensity of precipitation play an important role in triggering landslides. The precipitation data used in this study are obtained from the NASA Tropical Rainfall Measuring Mission (TRMM) Multi-Satellite Precipitation Analysis (TMPA) (Huffman et al. 2007), which provides calibrated sequential scheme for combining precipitation from multiple satellite and gauge analyses at a resolution of 0.25° × 0.25° over the latitude band 50°N–50°S every 3 h. The real-time rainfall is available on the NASA TRMM Web site (http://trmm.gsfc.nasa.gov). For the first case study in Indonesia, precipitation forcing also includes model simulations from Weather Research Forecasting (WRF) model forecasts provided at 4 km spatial resolution and 3-hourly temporal resolution. These types of precipitation products are used in the SLIDE model as dynamic forcing.

3 Application of SLIDE Model at Regional Scale

3.1 Case Study 1—Landslides Triggered by Typhoons in Indonesia

Indonesia suffers from typhoons, typhoon-triggered floods, and landslides during rainy seasons every year (Wardani and Kodoatie 2008). According to a landslide inventory database (Kirschbaum et al. 2009a, b), 402 people were killed by 17 landslides in 2003, and 243 people were killed by 13 landslides in 2007 in Indonesia. Among these landslides, 12 rainfall-induced landslides occurred in Java Island, see Fig. 2. During the period from 24th December 2007 to 2nd January 2008, the central parts of Java received close to 250 mm of rainfall. Storms delivering above average

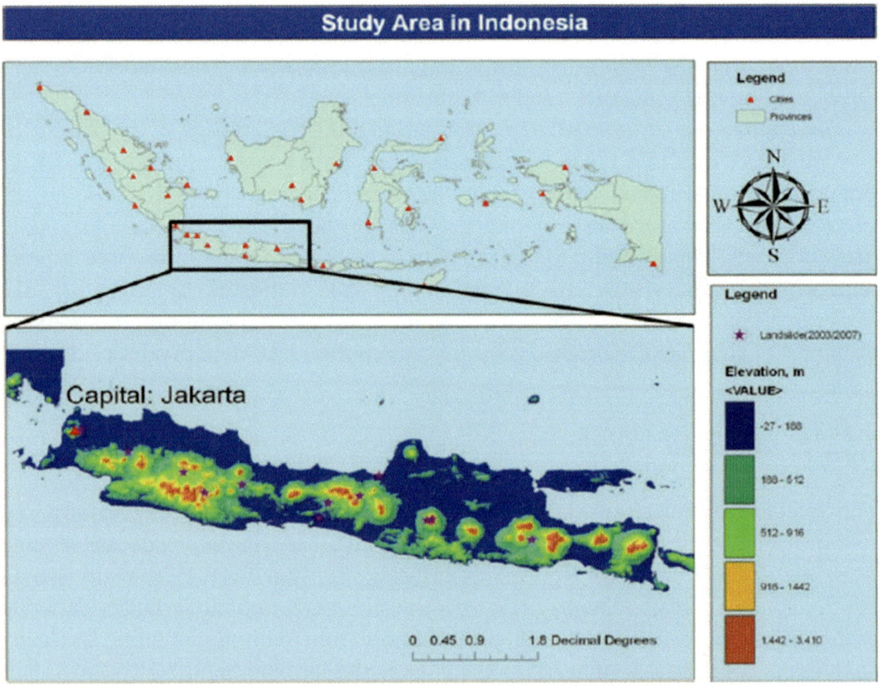

Fig. 2 The study area and the landslide inventory (2003, 2007) in Java Island, Indonesia

rainfall can create dangerous scenarios for regional flooding and slope failure in Karanganyar, Central Java. On 26th December 2007, two major landslides triggered by this typhoon-triggered rainstorm killed at least 65 people in Karanganyar and several other landslides occurred and claimed more lives according to the local media reports. SLIDE was implemented in Karanganyar, Java during the typhoon event to model one of its triggered landslide events (square B in Fig. 6, see Liao et al. 2010).

FS values are calculated for one landslide induced by rainstorms on 26th December 2007, in Karanganyar, Central Java using the SLIDE model. In order to reduce the computational load, soil depth is assumed to be 2 m instead of varying depth. The landslide has been simulated with TRMM and WRF precipitation data. A landslide warning would be issued when the FS value reaches below the critical value of 1, as expressed in Eq. (1). Figure 3a denotes a predicted slope failure at the 23rd time step (shown by the yellow arrow) in the TRMM rainfall series, where the FS value is below the critical value indicated by the dashed line. FS values decrease after rainfall reaches its peak value of approx. 50 mm over 3 h. The total rainfall keeps accumulating until it reaches above 150 mm. Although the prediction time shows a 3-h delay, it is still in agreement with the date of the event occurrence reported in local news on 26th December 2007. In addition to the TRMM rainfall, the model has been run by WRF precipitation to investigate how well the model would perform in the early-warning system. Figure 3b shows the WRF rainfall series with much lower intensity compared with TRMM for the same event over the Landslide Site B. The WRF rainfall gives a storm event accumulation of rainfall 150 mm, which agrees with the TRMM values. In the WRF-driven simulation, FS values show the same trend in time with the lowest value of 0.99 indicating the occurrence of a landslide.

3.2 *Case Study 2—Landslides Triggered by Hurricane Mitch in Honduras*

Intense rainfall from Hurricane Mitch from 27th to 31st October 1998 exceeded 900 mm within some regions of Honduras and triggered in excess of 50,000 landslides throughout the country (Harp et al. 2002). The hurricane and related landslides killed an estimated 5,657 people and injured another 12,272 according to the National Emergency Cabinet. Two study areas were selected to compute the FS maps with each covering approximately 600 km^2 over the south coast of Honduras (Fig. 4). In the selected study areas, the highest recorded rainfall intensity was 12 mm/h, with an accumulation close to 200 mm in 5 days. This region was selected for analysis because it currently has one of the largest and most regionally extensive landslide inventory data sets compiled following Hurricane Mitch.

Using TMPA data as the rainfall input, the parameterized SLIDE model was run at 3-h resolution and a FS value was generated throughout the Hurricane Mitch

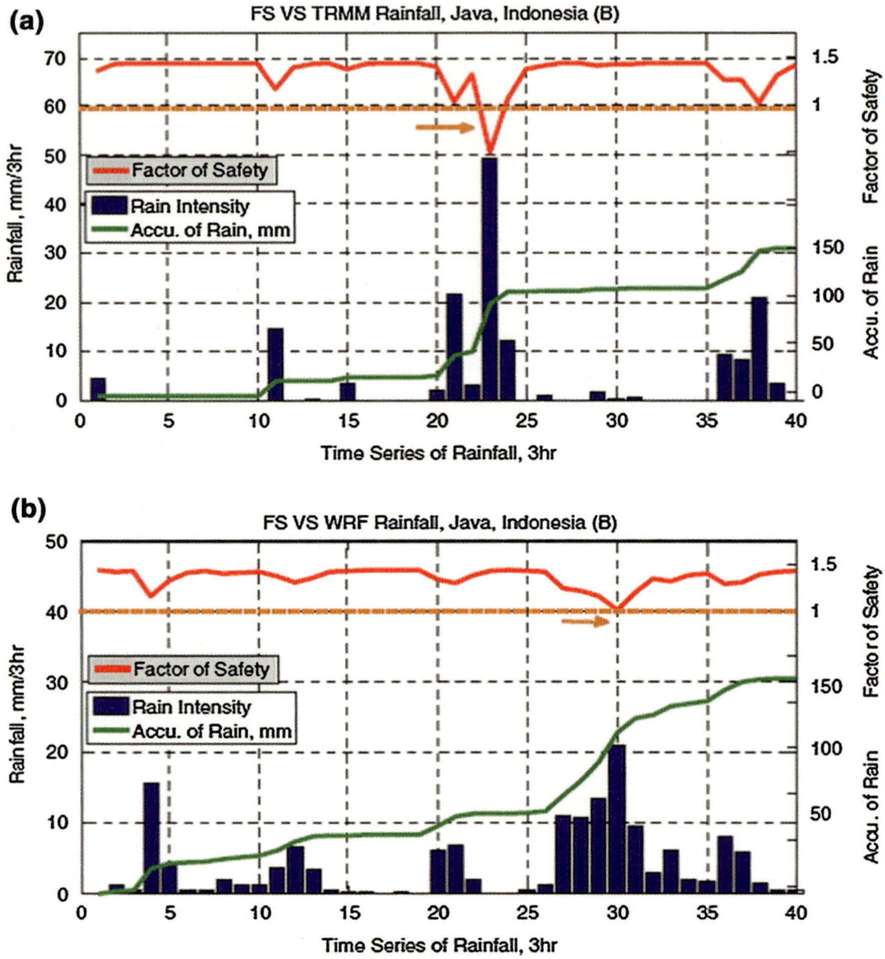

Fig. 3 **a** Time series of TRMM-driven FS and TRMM rainfall intensity during the time course of the landslide event B; and **b** the same as (**a**) but for WRF-driven FS and rainfall intensity

event from 25th November 1998 to 5th December 1998, at every 30 m grid cell. Figure 5 illustrates the unstable areas computed by the SLIDE model for the two study areas over the Hurricane Mitch event. Unstable areas (FS ≤ 1) are shown in red and observed landslides are shown in black.

Firstly, the results were evaluated by comparing the unstable maps with observation by four indices from Fawcett (2006). Labels {Y, N} were used for the class forecasts produced by the physical model in 30 m × 30 m cell grids. The labels $\{p, n\}$ were used for the class of observation in the field. In the class of observation, landslide areas were computed within a radius of 500 m. There are four possible outcomes when classifying a grid from the unstable map. If a computed unstable

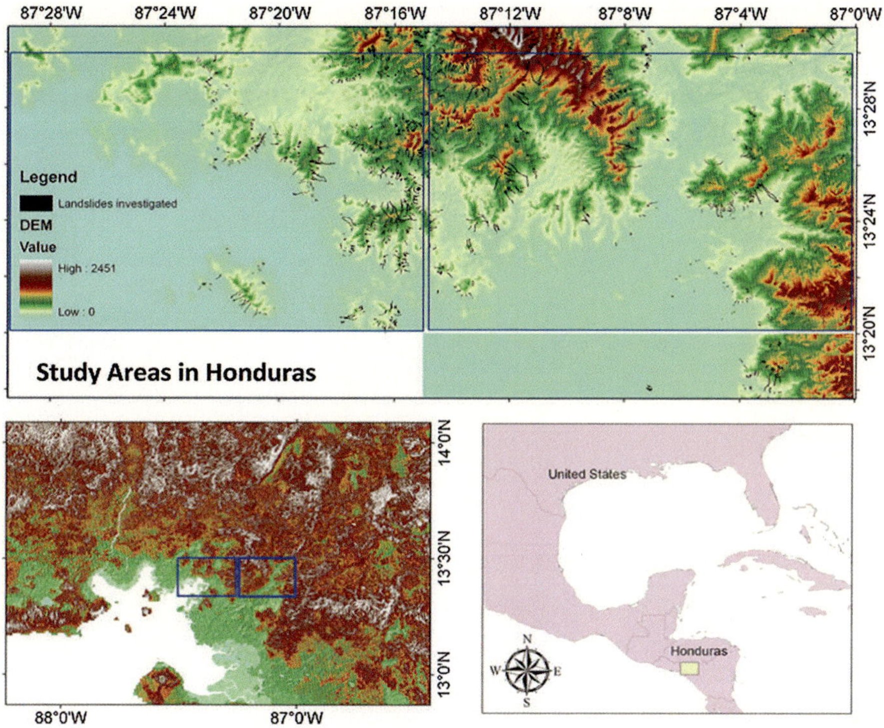

Fig. 4 Two study areas in Honduras for Hurricane Mitch in 1998 (*above banner*), zooming in from the Central America

grid is inside the observed landslide area, it is counted as *true positive* (TP, also called hit or rates of successful estimation of slope failure locations); if it is outside the observed landslide area, it is counted as *false positive* (FP, also called false alarm). If a computed stable grid matches an observed landslide grid, it is counted as *false negative* (FN); otherwise, it is called *true negative* (TN). The TP rate defines how well the predicted results agree with the observations. The TP rates are 78 % for the study area 'a' (Fig. 5a) and 75 % for the study area 'b' (Fig. 5b). The FP rate indicates the tradeoff between predicted results and observations. The FP rates are 1 and 6 % for study areas 'a' and 'b,' respectively. The error rate is defined as the portion of the computed unstable grids that did not contain observed landslides (Sorbino et al. 2010). The SLIDE model provides a value of error rate of 35 % in Region 'a' (Fig. 5a) while error rate is 49 % in Region 'b' (Fig. 5b). Accuracy represents the level of agreement between the forecast and the observations of landslides. The accuracy values are 98 % for the study area 'a' and 92 % for the study area 'b'. The precision means positive predictive value. The precisions are 65 and 51 % for study areas 'a' and 'b,' respectively (Fig. 5). All the statistics are listed in Table 2.

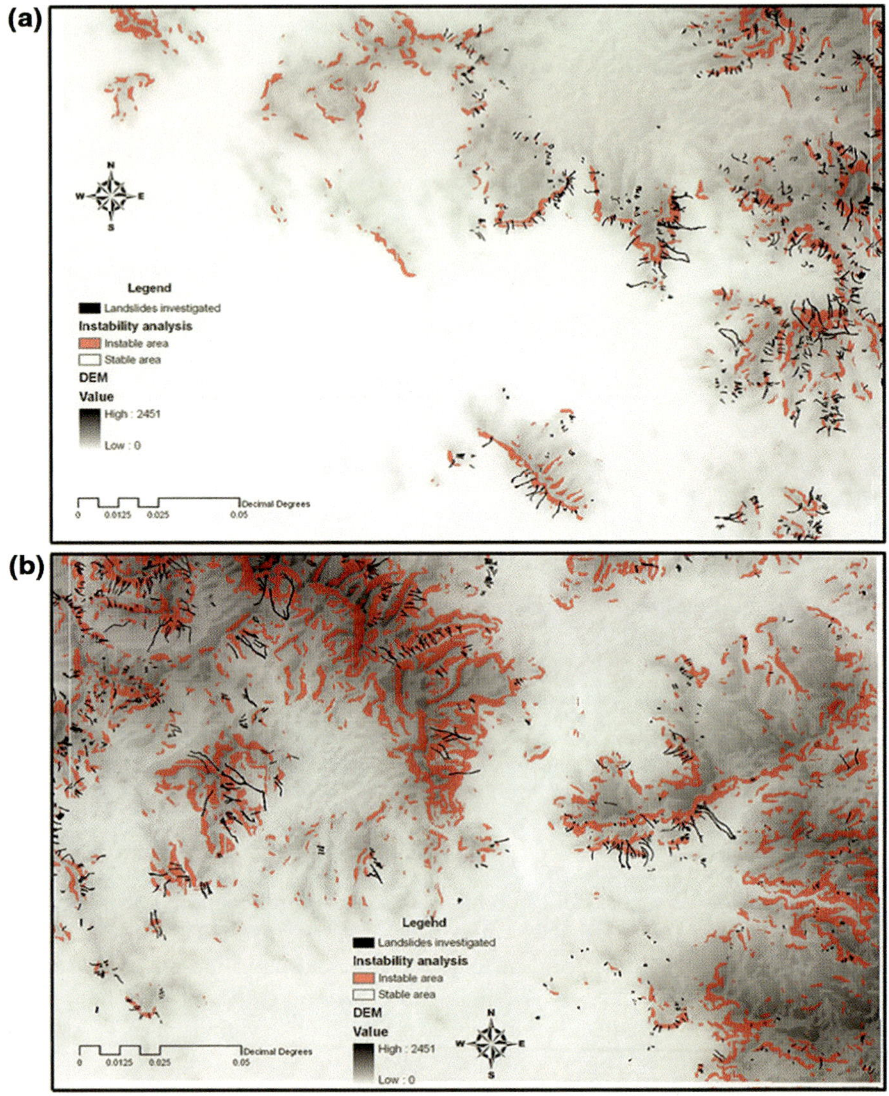

Fig. 5 Instability maps of landslides produced by SLIDE for the study areas (**a**, **b**)

The model forecasts show good agreement with the inventory over the two study areas. However, the model over-predicts the occurrence of landslides, particularly in the west part of the study area 'a' (Fig. 5a). In the study area 'b,' the regions with landslides predicted by the model are much larger than the reported areas. In general, the model forecast results in larger hazard zones than what is observed, resulting in an overestimation of hazard areas. This is consistent with other previous studies performed at regional scales, see Nadim et al. (2006).

Table 2 Evaluation statistics for instability maps of landslides produced by SLIDE

Indices	True positive rate (%)	False positive rate (%)	Accuracy (%)	Precision (%)	Error rate (%)
Study area (a)	78	1	98	65	35
Study area (b)	75	6	92	51	49
Perfect values	100	0	100	100	0

4 Summary and Conclusions

A physically based landslide forecast model SLIDE has been presented for assessment of shallow landslides triggered by typhoon rainstorm in Indonesia and Hurricane Mitch in Honduras. For the case study in Indonesia, results show that the SLIDE model successfully predicted the occurrence of the landslides on the dates of the real events. The prediction time from the modelling results shows only a 3-h delay on 26th December 2007. Regarding the forecast capabilities of SLIDE in Honduras, the high true positive rates and the low false positive rates indicate the effectiveness to employ SLIDE for the forecast of landslide disasters using remote sensing and geospatial data sets over larger regions. However, over-prediction is clearly shown by the error rates that are 35 and 49 % for study regions 'a' and 'b,' respectively. The over-sensitivity of the SLIDE model is possibly due to some simplified assumptions and limitations within this model. Overall, the agreement between the SLIDE modelling results and landslide observations in the two case studies demonstrates generally good predictive skills of SLIDE.

In the work presented here, the SLIDE model shows a potential for integration into regional-scale early-warning systems for landslides triggered by extreme precipitation events. Different from the framework based on empirical models, SLIDE can be used as a core to set up a physical framework to predict regional landslides by mapping the FS over a defined region, since it can link water content in the soil bulk and rainfall series more realistically. Although SLIDE has the potential to become a next-generation tool for early warning of landslides, it neglects some factors that can be important and may therefore affect the accuracy of model forecasting. For example, uniform geological structures of slopes are assumed as parameterization, which could lead to potential errors of modelling results. Besides this, it decreases the accuracy of modelling results by applying same soil values to the regional-scale modelling. This hypothesized procedure would also reduce the spatial variability. In addition, the simplified hydrological processes, which were made to make the model easily applicable over larger areas, may also influence the accuracy of simulations. Specifically, run-off and evapotranspiration are ignored in the water balance in SLIDE. All of the rainfall is assumed to infiltrate into the soil, which is not physically realistic. The deficiency of model physics requires making

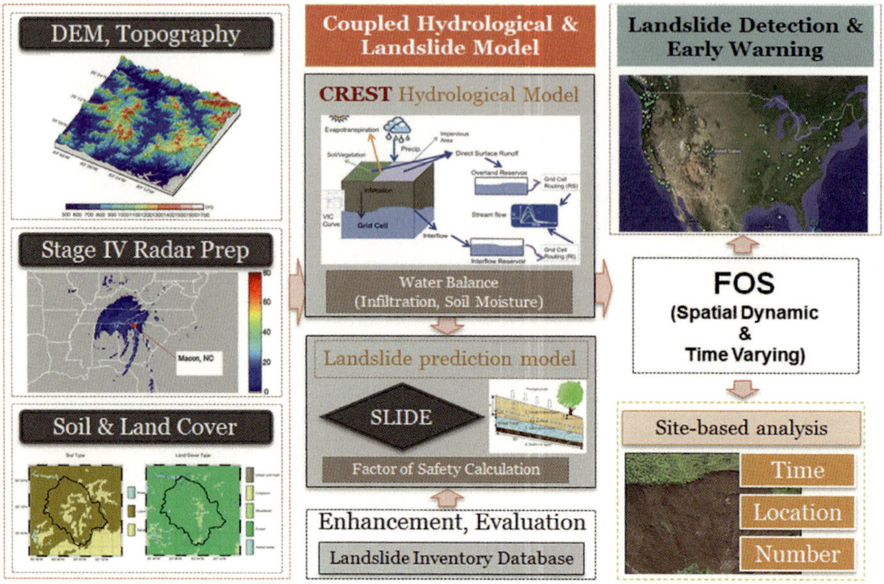

Fig. 6 Conceptual framework of the early-warning system through the coupled hydrological (CREST) and geotechnical model (SLIDE) for rainfall-triggered landslides

the coupling work between hydrological model (e.g., CREST model, see Wang et al. 2011) and geotechnical model (SLIDE) to have a better representation of the subsurface processes. We anticipate the development of future warning systems (Fig. 6) for rainfall-triggered landslides to be enhanced through this coupling strategy by considering the water balance (infiltration and soil moisture). Site-based analysis will provide valuable hints for us to gain a better and more comprehensive understanding of the model performance accuracy if additional landslide inventories, detailed soil information, and in situ measurements are available in the future.

Acknowledgments The authors of this chapter acknowledge the partial fund support from NASA, Oklahoma Department of Transportation, USA.

References

Baum, R. L., Savage, W. Z., & Godt, J. W. (2002). TRIGRS—a fortran program for transient rainfall infiltration and grid-based regional slope stability analysis. *United States Geological Survey Open File Report*.

Boebel, O., Kindermann, L., Klinck, H., Bornemann, H., Plotz, J., Steinhage, D., et al. (2006). Satellite remote sensing for global landslide monitoring. *EOS Transactions, American Geophysical Union, 87*(37), 357–358.

Dai, E. C., Lee, C. F., & Nagi, Y. Y. (2002). Landslide risk assessment and management: An overview. *Engineering Geology, 64*, 65–87.

Dietrich, W. E., Reiss, R., Hsu, M. L., & Montgomery, D. R. (1995). A process based model for colluvial soil depth and shallow landsliding using digital elevation data. *Hydrological Process, 9*, 383–400.

Fawcett, T. (2006). An introduction to ROC analysis. *Pattern Recognition Letters, 27*, 861–874.

Godt, J. W., Baum, R. L., & Lu, N. (2009). Landsliding in partially saturated materials. *Geophysical Research Letters 36*, paper no. L02403.

Harp, E. L., Hagaman, K. W., Held, M. D., & McKenna, J. P. (2002). Digital inventory of landslides and related deposits in Honduras Triggered by Hurricane Mitch. *United States Geological Survey Open-File Report 02-61.*

He, X., Hong, Y., Yu, X., Cerato, A. B., Zhang, X., & Komac, M. (2014). Landslides susceptibility mapping in Oklahoma state using GIS-based weighted linear combination method. In: K. Sassa, et al. (Eds.), *Landslide science for a safer geoenvironment* (Vol. 2, pp. 371–377). Heidelberg: Springer.

Hong, Y., Adler, R. F., & Huffman, G. (2006). Evaluation of the potential of NASA multi-satellite precipitation analysis in global landslide hazard assessment. *Geophysical Research Letters 33*, paper no. L22402.

Hong, Y., Adler, R. F., & Huffman, G. (2007). Use of satellite remote sensing data in mapping of global shallow landslides susceptibility. *Natural Hazards, 43*(2), 245–256.

Huffman, G. J., Adler, R. F., Bolvin, D. T., Gu, G., Nelkin, E. J., Bowman, K. P., et al. (2007). The TRMM multisatellite precipitation analysis (TMPA): Quasi-global, multiyear, combined-sensor precipitation estimates at fine scales. *Journal of Hydrometeorology, 8*, 38–55.

Iverson, R. M. (2000). Landslide triggering by rain infiltration. *Water Resources Research, 36*(7), 1897–1910.

Kirschbaum, D. B., Adler, R. F., Hong, Y., Hill, S., & Lerner-Lam, A. L. (2009a). A global landslide catalog for hazard applications: Method, results and limitations. *Natural Hazards, 52*(3), 561–575.

Kirschbaum, D. B., Adler, R. F., Hong, Y., & Lerner-Lam, A. L. (2009b). Evaluation of a preliminary satellite-based landslide hazard algorithm using global landslide inventories. *Natural Hazards and Earth System Science, 9*, 673–686.

Kirschbaum, D. B., Adler, R. F., Hong, Y., Peters-Lidard, C., & Lerner-Lam, A. L. (2012). Advances in landslide hazard forecasting: Evaluation of a global and regional modeling approach. *Environmental Earth Sciences, 66*, 1683–1696.

Liao, Z., Hong, Y., Wang, J., Fukuoka, H., Sassa, K., Karnawati, D., et al. (2010). Prototyping an experimental early warning system for rainfall-induced landslides in Indonesia using satellite remote sensing and geospatial datasets. *Landslides, 7*(3), 317–324.

Liao, Z., Hong, Y., Kirschbaum, D., & Liu, C. (2011). Assessment of shallow landslides from Hurricane Mitch in central America using a physically based model. *Environmental Earth Sciences, 66*(6), 1697–1705.

Lu, N., & Godt, J. W. (2008). Infinite slope stability under unsaturated seepage conditions. *Water Resources Research* 44, paper no. W11404.

Montrasio, L., & Valentino, R. (2008). A model for triggering mechanisms of shallow landslides. *Natural Hazards and Earth System Science, 8*, 1149–1159.

Nadim, F., Kjekstad, O., Peduzzi, P., Herold, C., & Jaedicke, C. (2006). Global landslide and avalanche hotspots. *Landslides, 3*, 159–173.

Sidle, R. C., & Ochiai, H. (2006). *Landslides: Processes, prediction, and land use*. In: Proceedings of AGU Conference 2006, Washington DC.

Sorbino, G., Sica, C., & Cascini, L. (2010). Susceptibility analysis of shallow landslides source areas using physically based models. *Natural Hazards, 53*, 313–332.

Taylor, D. W. (1948). *Fundamentals of soil mechanics*. New York: Wiley.

Wang, J., Hong, Y., Li, L., Gourley, J., Khan, S., Yilmaz, K., et al. (2011). The coupled routing and excess storage (CREST) distributed hydrological model. *Hydrological Sciences Journal, 56*(1), 84–98.

Wardani, S. P. R., & Kodoatie, R. J. (2008). Disaster management in Central Java Province, Indonesia. In: H. Liu & A. Deng, J. Chu (Eds.), *Geotechnical engineering for disaster mitigation and rehabilitation* (pp. 254–259). Heidelberg: Springer.

Wu, W., & Sidle, R. C. (1995). A distributed slope stability model for steep forested basins. *Water Resources Research, 31*, 2097–2110.

Part III
Satellite Remote-Sensing Techniques

Monitoring Landslide Activities in the Three Gorges Area with Multi-frequency Satellite SAR Data Sets

Lu Zhang, Mingsheng Liao, Timo Balz, Xuguo Shi and Yanan Jiang

Abstract Thousands of landslides are distributed along Yangtze River and its tributaries in the Three Gorges area from Chongqing Municipality in the west to Hubei Province in the east (P.R. China). Since the construction and regular operation of the Three Gorges Dam in the past two decades, many ancient landslides have been reactivated and some new landslides were formed along with the unprecedentedly huge water-level changes. Monitoring landslide activities has then been considered as a high-priority task for geological disaster prevention and management in the reservoir area, while traditional monitoring methods can hardly meet the requirements. In this chapter, we investigated the applications of several methods using Synthetic Aperture Radar (SAR) datasets in landslide monitoring in the Three Gorges area. Multifrequency satellite SAR data sets acquired by ENVISAT/ASAR, ALOS/PALSAR, and TerraSAR-X from different orbits were analyzed to retrieve historic deformations of a few typical landslides. The experimental results suggested that SAR Interferometry (InSAR) methods can be effectively used to monitor slow-moving landslides, while pixel offset tracking method is more suitable for detecting deformation of fast-moving landslides. Furthermore, qualitative correlation analyses indicated that variation of reservoir water level, particularly the rapid water-level decrease due to discharge, should be identified as a key driving factor for landslide deformation in the Three Gorges area.

L. Zhang (✉) · M. Liao · T. Balz · X. Shi · Y. Jiang
State Key Laboratory of Information Engineering in Surveying, Mapping and Remote Sensing, Wuhan University, Wuhan, China
e-mail: luzhang@whu.edu.cn

M. Liao
e-mail: liao@whu.edu.cn

T. Balz
e-mail: balz@whu.edu.cn

X. Shi
e-mail: xuguoshi@whu.edu.cn

Y. Jiang
e-mail: yananjiang@whu.edu.cn

M. Scaioni (ed.), *Modern Technologies for Landslide Monitoring and Prediction*, Springer Natural Hazards, DOI 10.1007/978-3-662-45931-7_10

Keywords Landslide • Deformation • SAR • Interferometry • Pixel offset tracking • Three gorges

1 Introduction

As the largest water conservancy project across the world, construction of the Three Gorges Project (TGP) was launched in 1994 and completed in 2009. During this period, water level in the Three Gorges Reservoir area rose stepwise up to 135, 156, and 175 m in 2003, 2006, and 2010, respectively, due to impoundment operations. Since 2010, the reservoir water level underwent an annual cycle of increasing to 175 m in early winter for hydroelectricity generation and decreasing to 145 m in early summer for flood prevention every year.

As an unprecedented complex and highly controversial project, debates on advantages and disadvantages of TGP have never stopped since its construction. As the largest power station in China, the power generated by TGP accounts for about 14 % of China's total hydroelectricity generation. TGP also improves the ability of flood control and navigation in Yangtze River Basin. On the other hand, tremendous human and economic resources have been invested in the TGP construction and problems also arose at the same time. Hundreds of towns including some historic relics submerged into the reservoir formed by the water-level rising; thus, thousands of people have to migrate to higher places or even to other provinces before and during the TGP construction, resulting in many social consequences. Furthermore, environmental problems such as water quality degradation and biodiversity disturbance have also appeared along with the rising water level.

Another important problem after the creation of the huge water basin is represented by the more active geological hazards, especially landslides. Steep slopes in the Three Gorges area are already at high risks for the occurrence of landslide before impoundment (Jackson and Sleigh 2000). The landslide stability was definitely affected by the rising water level. The number of active landslides after the tentative impoundment to water level of 175 m since 2008 is almost ten times of that in 1992–1995 (Cojean and Caï 2011). As a famous example, the Qianjiangping landslide located near the south bank of Yangtze River collapsed just a few days after the first impoundment in 2003 and caused huge losses of lives and properties (Wang et al. 2008). Other landslides such as Shuping in Zigui County and Huangtupo in Badong County are also in unstable status (Wang et al. 2008; Liu et al. 2013).

Landslides pose great threats to lives and properties of local residents. Therefore, landslide deformation monitoring became a vital task for disaster prevention and alert. The Chinese government have spent budget of billions RMB to stabilize landslides in the Three Gorges area (Plateau 2006). Landslide monitoring methods, such as geodetic control networks based on global navigation satellite system (GNSS—see Hofmann-Wellenhof et al. 2007; Costantino and Angelini 2011; Pirotti et al. 2014), optical leveling (Fastellini et al. 2011; Pingue et al. 2013), optical remote sensing (Delacourt et al. 2007; Debella-Gilo and Kääb 2012; Travelletti et al. 2012; Barazzetti et al. 2014), as well as geotechnical sensors (Angeli et al. 2000), have been largely used in landslide monitoring (see some examples in Wang et al. (2008), Li et al. (2010), Miao et al. (2014)). Advantages and disadvantages of these methods are summarized in Table 1.

Major problems for geodetic leveling, GNSS, and geotechnical sensors are that only spatially sparse observations can be obtained and it is difficult to employ these methods on steep slopes due to high cost and inaccessibility. Optical remote sensing known for its capability of wide coverage has been applied in landslide risk assessment and mapping (Liu et al. 2004; Nichol et al. 2006; Fourniadis et al. 2007). However, its application is usually limited by bad weather conditions such as cloud, fog, smoke, rain, and snow.

By contrast, *Synthetic Aperture Radar* (SAR) is an active imaging system with unique ability of acquiring data over large areas independent of solar illumination and weather conditions. SAR Interferometry (InSAR) technology can be used to measure subtle ground displacements by exploitation of phase difference information. It has been successfully applied in geohazard mapping such as landslides (Fruneau et al. 1996; Rizo and Tesauro 2000; Berardino et al. 2003; Xia et al. 2004), earthquakes Fornaro et al. (2012), and ground subsidence Perissin and Wang (2011).

Nevertheless, temporal/spatial de-correlations and atmospheric phase screen (APS) constitute major factors hindering applications of traditional InSAR methods (Nagler et al. 2002; Xia et al. 2004). Careful selection of suitable data pairs in consideration of normal baseline and temporal baseline should be performed to mitigate unfavorable impacts of these factors. The advanced differential InSAR methods (A-DInSAR—Crosetto et al. 2005), namely Permanent/Persistent Scatterer InSAR (PSI) (Ferretti et al. 2000, 2001; Hooper et al. 2004) and small baseline

Table 1 Comparison of different methods for landslide monitoring

Methods	Advantages	Disadvantages
Optical leveling	Mature technology, high precision, reliable information, convenient and flexible	High cost, small coverage, low density, limited by steep topography, no continuous observation
Global Navigation Satellite System (GNSS)	All-weather surveillance, high precision, convenient and flexible, continuous observation	High cost, low density, limited by steep topography and multipath effects
Optical remote sensing	High degree of automation, large coverage, low cost	Low accuracy, suitable for landslides with large displacement, limited by clouds and water vapor
Geotechnical sensors	High precision	High cost, in situ installation on landslides required
SAR interferometry (InSAR)	Large coverage, low cost, high precision	Temporal/spatial de-correlations and atmospheric phase screen

subset (SBAS) (Berardino et al. 2002; Lanari et al. 2004), were proposed to overcome these limitations through time series analyses of repeat-pass SAR images. Advanced DInSAR methods were successfully applied in slow-moving landslide mapping (Hilley et al. 2004; Cascini et al. 2009, 2010; Handwerger et al. 2013; Herrera et al. 2013). But for fast-moving landslides, underestimation often occurs in areas with dense vegetation and steep slopes even using the advanced DInSAR methods, especially for X-band SAR data of short wavelength.

Different from A-DInSAR technology that depends on phase observation, non-interferometric *pixel offset tracking* (POT), or *amplitude tracking* method (Raucoules et al. 2013) makes use of only amplitude information to measure displacements through subpixel-level SAR image matching. Although the precision of POT measurement is usually lower than that of InSAR, POT is much more robust to de-correlation and APS problems and thus more suitable for mapping large displacements caused by earthquakes, landslides, and glacier motion (Schubert et al. 2013). The precision can reach 1/20 pixels if high amplitude correlation is guaranteed, but such high precision is generally difficult to achieve in vegetated areas. The matching errors in vegetated areas are usually high, and they will degrade the global displacement measurements. Pointlike target offset tracking (PTOT) method is an improvement of the traditional pixel offset method (Hu et al. 2014). By making use of stable bright targets such as artificial *corner reflectors* (CR's) and buildings in SAR images, this method can be used as a viable alternative approach for difficult areas where A-DInSAR methods are difficult to be applied.

Since landslide problem is a big concern in the Three Gorges area, monitoring the movement of slopes in this area has also become a hot topic. Several different methods were used in previous studies. GPS data from 21 landslides were used to analyze mechanism of the slow-moving landslides in Jurassic red strata of Three Gorges Reservoir (Miao et al. 2014). A real-time monitoring and early-warning system combining GPS receivers, boreholes, inclinometers, and piezometers was employed on landslides at the relocated Wushan town in the Three Gorges Reservoir area (Yin et al. 2010). Extensometer measurements combined with GPS data were used for monitoring the deformation of Shuping landslide (Wang et al. 2008). Remotely-sensed images combined with field data were used for landslide hazard assessment (Liu et al. 2004; Fourniadis et al. 2007). DInSAR was firstly used for landslide monitoring in Three Gorges area combined with CR's, and promising results were achieved (Xia et al. 2002, 2004). In recent years, A-InSAR methods with *time series* (or *stacks*) of SAR data were also successfully applied for landslide monitoring in Badong and Zigui areas (Liao et al. 2012; Perissin and Wang 2012; Liu et al. 2013; Tantianuparp et al. 2013). InSAR is a useful tool for slow-moving landslide monitoring, while for fast-moving landslide such as the Shuping landslide, underestimation by InSAR measurements was identified when compared with extensometer measurements. As an alternative solution, pixel offset tracking method with SAR images were successfully used in deformation measurement (Li et al. 2011).

In this chapter, basic principles of deformation mapping methods using SAR images are first introduced in Sect. 2. Then, our test areas and data sets used are described in Sect. 3. Experimental results in Badong and Zigui areas are presented in Sect. 4. Finally, discussion and conclusion are given in Sect. 5.

2 Methodology

2.1 Differential Interferometry

Figure 1 shows the geometry of InSAR. T is a point target on the Earth surface with height h relative to the ellipsoid (Schofield and Breach 2007). Point T was observed by the radar at two separate times when the satellite was at positions P_1 and P_2, respectively. H is the height of satellite at position P_1. The distances from the satellites at P_1 and P_2 to the target T are ρ_1 and ρ_2, respectively. θ is the look angle of satellite at P_1. The distance between P_1 and P_2 is defined as *spatial baseline B*. α is the angle between the spatial baseline P_1P_2 and the ground-range direction. $B_{\|}$ is the baseline component parallel to the line-of-sight (LOS) direction P_2T. $B_{\perp}$ is the baseline component perpendicular to the LOS direction. B_h and B_v are the horizontal and vertical components of the baseline, respectively. The relationship between $B_{\|}$, $B_{\perp}$, B_h, and B_v can be expressed as follows:

$$B_{\|} = \rho_2 - \rho_1 = B\sin(\theta - \alpha) \tag{1}$$

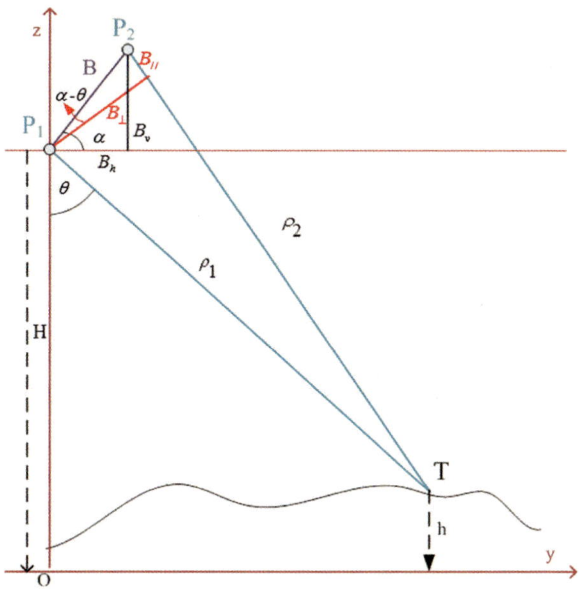

Fig. 1 Imaging geometry of SAR interferometry

$$B_{\perp} = B\cos(\theta - \alpha) = B_h \cos\theta + B_v \sin\theta \tag{2}$$

The phase of backscattered signal from the Earth surface is determined by the distance from the satellite to the target and scattering properties, which can be expressed as follows:

$$\phi = \phi_r + \phi_{\text{scat}} \tag{3}$$

where ϕ_r and ϕ_{scat} depend upon the distance from satellite to targets and the scatter properties of the targets in the resolution cell. ϕ_r can be expressed as a function of the distance between satellite and target:

$$\phi_r = -\frac{4\pi}{\lambda} \cdot \rho \tag{4}$$

where λ is wavelength of radar signal. The minus symbol represents the phase delay.

The phase difference between SAR images of two separate observations on the same target is named as the interferometric phase ϕ_{IF}:

$$\begin{aligned}\phi_{\text{IF}} &= \phi_2 - \phi_1 \\ &= -\frac{2\pi}{\lambda} k(\rho_2 - \rho_1) + \left(\phi_{\text{scat},2} - \phi_{\text{scat},1}\right) \\ &= -\frac{2\pi}{\lambda} \Delta\rho + \Delta\phi_{\text{scat}}\end{aligned} \tag{5}$$

where $\Delta\rho = \rho_1 - \rho_2$ is the distance difference between the two acquisitions caused by the height and deformation in the LOS direction, while $\Delta\phi_{\text{scat}}$ is determined by the scattering properties in the resolution cell. The scattering properties change with time and different incidence angle. Assuming that ϕ_{scat} keeps unchanged in two SAR images, which means $\Delta\phi_{\text{scat}} = 0$, the interferometric phase can be expressed as follows:

$$\begin{aligned}\phi_{\text{IF}} &= \phi_2 - \phi_1 - \phi_R \\ &= -\frac{2\pi}{\lambda} k(\rho_2 - \rho_1) - \phi_R \\ &= -\frac{2\pi}{\lambda} kB_{||} - \phi_R\end{aligned} \tag{6}$$

ϕ_R is the interferometric phase caused by distance difference in the line-of-sight (LOS) direction which is also called *flattened Earth phase*. The interferometric phase ϕ_{IF} can then be expressed as a function of height after removal of the flatten earth phase. Its derivative with respect to θ is as follows:

$$\frac{d\varphi_{\text{IF}}}{d\theta} = -\frac{2\pi}{\lambda} kB\cos(\theta - \alpha) = -\frac{2\pi}{\lambda} kB_{\perp} \tag{7}$$

The height of the target T is as follows:

$$h = H - \rho_1 \cos\theta \tag{8}$$

The derivative of height with respect to θ is as follows:

$$\frac{\mathrm{d}h}{\mathrm{d}\theta} = \rho_1 \sin\theta \tag{9}$$

From (7) and (9), we can derive

$$\begin{aligned} \frac{\mathrm{d}h}{\mathrm{d}\phi_{\mathrm{IF}}} &= \frac{\mathrm{d}h}{\mathrm{d}\theta} \cdot \frac{\mathrm{d}\theta}{\mathrm{d}\phi_{\mathrm{IF}}} = -\frac{\lambda \cdot \rho_1 \sin\theta}{2\pi \cdot k \cdot B_{\perp}} \\ &= -\frac{\lambda \cdot \rho_1 \sin\theta}{2\pi \cdot k \cdot (B_h \cos\theta + B_v \sin\theta)} \end{aligned} \tag{10}$$

The discrete form of (10) can be expressed as follows:

$$\begin{aligned} \Delta h &= -\frac{\lambda \cdot \rho_1 \sin\theta}{2\pi \cdot k \cdot B_{\perp}} \Delta\phi_{\mathrm{IF}} \\ &= -\frac{\lambda \cdot \rho_1 \sin\theta}{2\pi \cdot k \cdot (B_h \cos\theta + B_v \sin\theta)} \Delta\phi_{\mathrm{IF}} \end{aligned} \tag{11}$$

From Eq. (4), we can see that the change of interferometric phase ϕ_{IF} is linearly correlated with the change of target height. Therefore, the interferometric phase can be converted to the relative height of the target.

If the ground target moved during the interval between two observations, we call it deformation at this target. To measure the deformation signal, the phase corresponding to the target height needs to be removed and this procedure is called differential interferometry.

The differential phase $\Delta\phi_{\mathrm{diff}}$ can be expressed as follows:

$$\phi_{\mathrm{diff}} = \phi_{\mathrm{def}} + \Delta\phi_{\mathrm{topo}} + \phi_{\mathrm{atm}} + \phi_n \tag{12}$$

Meanings of the above four components are explained as below:

- *Residual topography phase* $\Delta\phi_{\mathrm{topo}}$: Although the topography phase has been removed using external DEM, there are still residual topography phase due to DEM uncertainty;
- *Deformation phase* ϕ_{def}: If the ground target has moved between both observations, the movement of the target results in a phase changes along the LOS direction. This deformation phase is usually correlated in time and space. The correlation pattern will be different according to the type of deformation;
- *Atmospheric phase* ϕ_{atm}: The atmospheric condition changes with time; thus, the phase delay caused by the atmosphere phase screen may be different in

repeat-pass InSAR. Atmospheric phase is correlated in space and uncorrelated in time; and

- *Noise phase* ϕ_n: The de-correlated noise is neither correlated in space and time. It can be random noise caused by the thermal noise and data processing errors.

2.2 Time Series Analysis

There are primarily two categories of A-DInSAR analysis methods, namely PSI and SBAS InSAR. They are based on two different basic principles. For PSI, pointlike targets with stable scattering behavior are first identified, and then, interferometric phase analyses are carried out on these targets only to ensure the reliability of deformation measurement results. For SBAS, interferograms with short temporal and normal baselines are grouped into multiple subsets to achieve high coherence. Singular value decomposition (SVD) method is then employed to extract deformation signal on pointlike targets detected from the subsets in case of rank defect.

2.2.1 Permanent/Persistent/Coherent Scatterer InSAR

Permanent/Persistent Scatterer InSAR (PSI) was first proposed by Politecnico di Milano (Italy) around 2000 (Ferretti et al. 2000, 2001). Indeed, the Permanent Scatterers® technique is patented by T.R.E. company, a spin-off of Politecnico di Milano University. In the following years, other similar solutions have been developed by other authors (Hooper et al. 2004; Perissin and Wang 2012). For this reason, the general term Persistent Scatterer Interferometry is preferred when referring to such solutions.

The basic concept of PSI is to identify a set of *Permanent Scatterers*® whose backscatter signals are stable over the whole time span covered by a long stack of SAR images. These points, which have been also called *Persistent Scatterers* to be independent from T.R.E. patented solution, will be only minimally affected by spatial/temporal de-correlation. In the following, we will use the acronym *coherent targets* (CT's) as recently proposed in Wasowki and Bovenga (2014). The backscatter signal of a CT is very high, and its size is usually smaller than a resolution cell. This property allows us to precisely locate the CT's on objects like man-made structures, bare rocks, and artificial CR's. Since CTs are almost not affected by temporal de-correlation, the *coherence* indicating the similarity of interferometric SAR echoes is very high. Models can be applied to CT's to separate the residual topography phase, the atmospheric phase, and the deformation phase. Readers may refer to (Ferretti et al. 2000, 2001, 2007) for more detail of about PSI, while in Wasowsky and Bovenga (2014), a review on the state-of-the-art applications for landslide investigation is reported.

A 'master' image should be firstly selected from the SAR data sets in consideration of spatial–temporal baselines and Doppler centroid frequency. All the 'slave' images are registered and resampled with respect to the master image.

Then, interferograms are formed by conjugate multiplication between the 'master' and resampled 'slave' images. Differential interferograms can be generated after removal of topographic phase assisted by a reference DEM.

Amplitude dispersion index is useful in selecting the CT candidates instead of *phase stability analysis* (Ferretti et al. 2001). Since the incidence angle and atmospheric condition varied with different acquisition times, the amplitude correspondent to the backscattered signal was also different. Calibrations to the amplitude should be done before the selection of candidate CT's. The amplitude dispersion value D_A can be defined as follows:

$$D_A = \frac{\sigma_A}{\mu_A} \tag{13}$$

where σ_A and μ_A are the standard deviation and mean of time series amplitude values, respectively. D_A is very close to the phase deviation if a high signal-to-noise ratio (SNR) is guaranteed. A sparse grid can be constructed using the extracted candidate CT's. The phase difference between neighboring candidates can be expressed as follows:

$$\begin{aligned}
\Delta\phi_{\text{diff}} &= \Delta\phi_{\text{defo}} + \Delta\phi_{\text{DEM-error}} + \Delta\phi_{\text{atmos}} + \Delta\phi_{\text{noise}} \\
&= \frac{4\pi}{\lambda}(T\Delta v + \Delta D_{\text{non-linear}}) + \frac{4\pi}{\lambda}\frac{B_\perp}{\rho \sin\theta}\Delta h + \Delta\phi_{\text{atmos}} + \Delta\phi_{\text{noise}} \\
&= \frac{4\pi}{\lambda}T\Delta v + \frac{4\pi}{\lambda}\frac{B_\perp}{\rho \sin\theta}\Delta h + \left(\frac{4\pi}{\lambda}\Delta D_{\text{non-linear}} + \Delta\phi_{\text{atmos}} + \Delta\phi_{\text{noise}}\right)
\end{aligned} \tag{14}$$

$$e = \frac{4\pi}{\lambda}\Delta D_{\text{non-linear}} + \Delta\phi_{\text{atmos}} + \Delta\phi_{\text{noise}} \tag{15}$$

where e is called phase residual. When $|e|$ is less than π, spatially phase unwrapping can be carried out. PSI takes the Δh and Δv as the estimated linear deformation and DEM error difference when the coherence γ reaches maximum:

$$|\gamma| = \left|\frac{1}{M}\sum_{k=1}^{M}\exp(je)\right| \tag{16}$$

Since we have estimated the linear deformation rate and the DEM error in the last step, these two terms can be then removed from the original differential phase, and then, we got the phase residual including nonlinear deformation phase, atmospheric phase, and noise. According to Hanssen (2001), the atmospheric signal is correlated in less than 1 km and behaves as spatially low-frequency signal. For a given pixel, the atmospheric phase can be regarded as random signal in time series. But for nonlinear deformation, its phase signal is both spatially and temporally correlated and shows up as a temporal low-frequency signal. Thus, we can separate the nonlinear deformation signal and the atmospheric signal by using spatial and temporal filters.

After we estimated the APS and removed it from the differential phase, the coherence of each candidate CT can be recalculated and those with coherence lower than a threshold will be rejected in the end. Then, phase unwrapping can be carried out on the final CT's to retrieve a precise estimation of the deformation component.

2.2.2 Small Baseline Subset

SBAS is another time series InSAR technique proposed to overcome de-correlation problems (Berardino et al. 2002; Lanari et al. 2004). Different from PSI that is based on a single master image, SBAS makes full use of interferograms with short temporal and spatial baselines with respect to multiple master images. On the other hand, differential interferograms after phase unwrapping were used to obtain the mean velocity map and the time series by SVD (Berardino et al. 2002).

Assuming that $N + 1$ SAR images have been acquired from time t_0 to time t_N and that one image should be selected first to serve as master, all the other images have to be coregistered and resample to the master image. M differential interferograms are formed and phase unwrapped. All the pixels in the unwrapped differential interferograms are calibrated with respect to a stable point. Pointlike targets with high coherence values were selected as candidates.

The interferometric phase for a given pixel in the differential interferograms formed by master image and slave image acquired at t_B and t_A ($t_B < t_A$) can be expressed as follows:

$$\phi_{\text{diff}} = \Delta T v + \Delta\phi_{\text{nl}} + \Delta\phi_{\text{topo}} + \Delta\phi_{\text{atm}} + \Delta\phi_n \tag{17}$$

where v is the linear deformation rate, $\Delta T = t_A - t_B$ is the temporal baseline, and $\Delta\phi_{\text{nl}}$ are the phase correlated with the nonlinear deformation. $\Delta\phi_{\text{topo}}$ is the topographic phase caused by the DEM uncertainty, $\Delta\phi_{\text{atm}}$ is the atmospheric phase, and $\Delta\phi_n$ is the noise term.

Since there are linear correlations between DEM uncertainty, topographic phase, linear deformation rate, and temporal baseline, usually a two-dimensional regression is used to retrieve the DEM error and the linear deformation rate (Shi et al. 2014).

After we have obtained an initial estimation of the DEM uncertainty and the linear deformation rate, these two components can be removed from the differential phase and the residual can be reformulated as follows:

$$\delta\phi_{\text{res}} = \Delta\phi_{\text{diff}} - \Delta\phi_{\text{topo}} - t \cdot v = \Delta\phi_{\text{nl}} + \Delta\phi_{\text{atm}} + \Delta\phi_n \tag{18}$$

Then, we can focus on the retrieval of nonlinear deformation series from the phase residuals. To separate the atmospheric phase and the nonlinear deformation phase, a low-pass filter in space and a high-pass filter in time can be applied on the residuals to isolate the atmospheric phase. And then, atmospheric phase can be removed. The residuals phase could be expressed as follows:

$$\delta\phi_{\text{res}} = \Delta\phi_{\text{diff}} - \Delta\phi_{\text{topo}} - t \cdot v_{\text{nl}} - \Delta\phi_{\text{atm}} = Bv_{\text{nl}} + \delta\phi_n \tag{19}$$

where v_{nl} is a generic element of the nonlinear deformation vector $v_{\text{l}} = [v_1, \ldots, v_N]$ and matrix **B** is a matrix of size $M \times N$, whose generic element (m, k) is defined as $B(m, k) = t_{k+1} - t_k$ for $i + 1 \leq k \leq j$, $\forall m = 1, \ldots, M_D$ and $B(m, k) = 0$ elsewhere (i and j are the indexes of the master image and the generic slave image, respectively, being $0 \leq i < j \leq N$). Then, nonlinear component of the deformation rate was retrieved by SVD in case that matrix **B** has full rank. The deformation rate of all the points is set as zero at the first acquisition date. Then, accumulated deformation could be computed by adding linear and nonlinear components together.

2.3 Point-like Target Offset Tracking

Pixel offset tracking method has been widely used for deformation mapping of earthquakes (Casu et al. 2011; Hu et al. 2014), ice flow (Schubert et al. 2013), and other natural processes. This technique makes use of only amplitude information of SAR images to measure displacements of subpixel precision by means of image matching techniques (Grün 2012). As we mentioned, InSAR method can only measure one-dimensional deformation in the LOS direction, and thus, it is much more sensitive to displacements in east–west and vertical directions than those in north–south direction. Different from InSAR method, pixel offset method can measure two-dimensional deformations in both azimuth and LOS directions, which means that this technique can also measure displacements in the north–south direction. Usually, offset measurements are of low accuracy in vegetated areas where the amplitude is relatively low. PTOT is an improvement of the traditional method through the use of those pixels featuring high amplitude values (Hu et al. 2014). There are mainly two major steps in the PTOT.

The first step is *point-like target selection*. Since the signal backscattered from an ideal pointlike target (PT) behaves as a 2D sinc function, we can calculate the cross-correlation between a moving window in the master image and an ideal sinc-like 2D impulse response function. Pixels with normalized correlation coefficients lower than 0.2 are excluded in the next processing step. Then, a pixel could be identified as a PT candidate if the product of the correlation coefficient and the amplitude is above an adaptive threshold, which is determined by the mean amplitude plus 1.5 times the standard deviation (std.dev.) in the moving window. Afterward, the locations of detected PT's are used as the input for the pixel-offset tracking. The second step is *pointlike target matching*. Offset tracking at PT's is similar to the standard method, i.e., cross-correlation between patches extracted around the master and the slave centers are performed. Both azimuth and range offsets measured consist of two components: (1) offsets induced by different orbits of master and slave images and (2) offsets caused by displacements happened during the acquisition interval. Usually, the orbital ramp can be separated from the initial offset by using a globally quadratic polynomial function, subject to the assumption that only a small part of the image is affected by the deformation process. A culling procedure is applied to fit the orbital ramp to avoid the biased polynomial coefficients induced by outliers of large displacements. Finally, pixel offsets caused by displacements can be obtained after removal of orbital offsets.

3 Study Area and Data Sets

3.1 Study Area

The Three Gorges Reservoir mainly extends from Chongqing Municipality to Hubei Province (P.R. China). Our study areas are located in Badong County and Zigui County of Hubei Province, where the tectonic transition zone from the second to third topographic step of China passes. The elevation ranges from 0 to 2,000 m a.s.l. There are many N–S-oriented high mountains with steep slopes. The NNE–SSW-oriented Huangling anticline mainly consists of pre-Sinian metamorphic and magmatic rocks which have high compressive resistance. Huangling anticline is confirmed to have a steeper western limb and a gentler eastern limb (Mei et al. 2013). The geological environment is favorable for dam construction, and the Three Gorges Dam was finally built in the town of Sandongping at Yichang city of Hubei Province.

A few faults are distributed in the study area, including Jiuwanxi fault and Xiannushan fault near the dam, and Niukou–Xiangluping fault located in the intersection area of Badong and Zigui. Other geological structures such as Guandukou syncline and Zigui syncline are shown in Fig. 2. A Mw 5.1 earthquake occurred around Badong County on 16th December 2013, shortly after the reservoir storage to the highest water level of 175 m. Seismic activities concerned with fault movements and water-level fluctuations have also happened in this area. Research showed that a relatively high rate of reservoir-induced seismicity was correlated with the increase of water level (Mei et al. 2013).

Landslides are a big concern in this area. The number of active landslides has rapidly increased since the construction of the Three Gorges Dam. According to Cojean and Caï (2011), 85 % of the unstable slopes are made up of Triassic or Jurassic formations. The strata are called Badong formation (mainly composed of sedimentary rocks of Triassic and Jurassic age) and widely distributed in Three Gorges area, and therefore, numerous landslides were developed in this area.

3.2 Test Data Sets

Since the first spaceborne SAR sensor was successfully launched in 1978 (Hanssen 2001), many countries and international agencies have developed their first-generation SAR satellites working in different frequencies: ERS-1/2 (1991 and 1995; C-band) and ENVISAT/ASAR (2002; C-band) launched by European Space

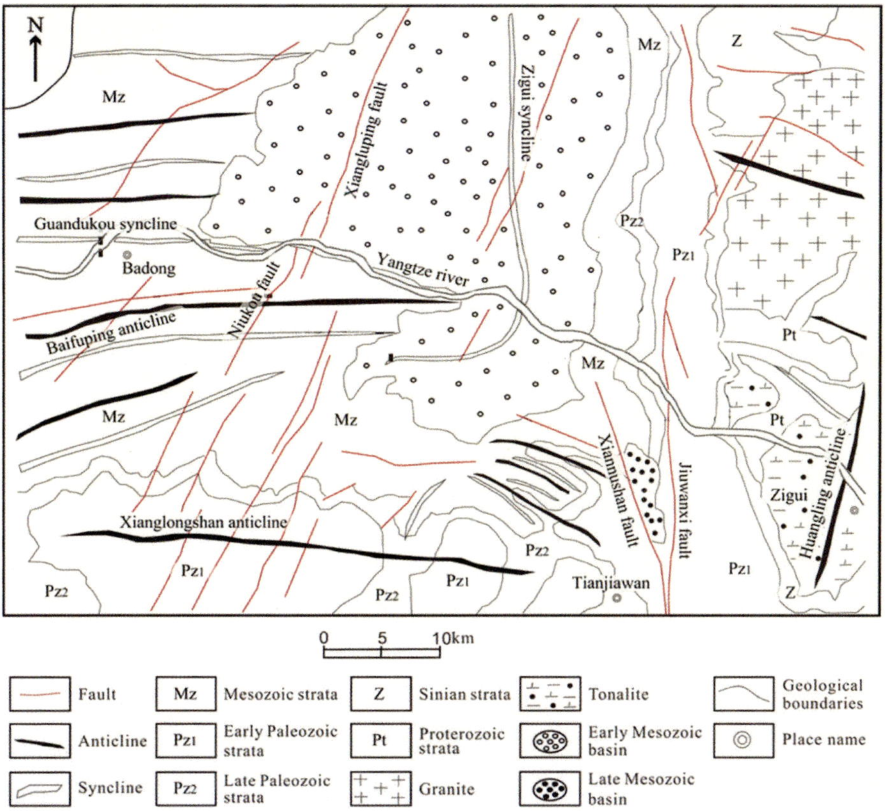

Fig. 2 Geological setting of the study area (reproduced with permission from Peng et al. 2014)

Agency (ESA); JERS-1 (1992; L-band) launched by Japan Aerospace Exploration Agency (JAXA); SIR-C/X-SAR (1994; C/X-band) launched by National Aeronautics and Space Administration (NASA) of the USA; and RADARSAT-1 (1995; C-band) launched by Canadian Space Agency (CSA). Series of breakthroughs have been made with the data collected by them. The second-generation satellite SAR systems have been developed and put into operation, including ALOS/PALSAR (2006; L-band) launched by JAXA, COSMO-SkyMed (2007, 2008 and 2010; X-band) launched by Italian Space Agency (ASI), TerraSAR-X/TanDEM-X (2007 and 2010; X-band) launched by German Aerospace Center (DLR), RADARSAT-2 (2007; C-band) launched by CSA, Sentinel-1 (2014; C-band) launched by ESA, and ALOS-2 (2014; L-band) launched by JAXA. Huge amount of data have been acquired and can provide us valuable information especially in inaccessible areas. A complete up-to-date review of missions operated and scheduled can be found in Wasowski and Bovenga (2014).

We have collected 22 images acquired by ALOS/PALSAR and two stacks of images acquired by ENVISAT/ASAR system, i.e., one stack of 17 images in

Table 2 Basic information of SAR data sets

	ENVISAT/ASAR		ALOS/PALSAR	TerraSAR-X
Orbit	Ascending	Descending	Ascending	Descending
Imaging mode	IMS		FBS/FBD	SSC
Acquisition time coverage	2006/02/11–2010/02/20	2008/03/23–2010/08/15	2006/12/27–2011/02/22	2008/08/12–2010/05/01
Central frequency	5.33 GHz		1.27 GHz	9.65 GHz
Polarization	VV		HH	VV
Incidence angle	23°		34.3°	24°
Pulse repetition frequency (PRF)	1,652 Hz		2,160 Hz	3,610 Hz
Ground resolution (azimuth × range)	30 m × 30 m		4.5 m × 7 m	3 m × 3 m

ascending orbit and the other stack of 21 images in descending orbit. This data sets was focused to study the Huangtupo landslide in Badong County. Thirty-four high-resolution TerraSAR-X images acquired from August 2008 to May 2010 were used to investigate the landslide activities in Zigui area. All the data sets were acquired in stripmap mode. Basic information of these data sets is summarized in Table 2.

As mentioned before, reference digital elevation models (DEM's) must be used to facilitate the removal of topographic phase during DInSAR analyses. Up to now, a few global coverage DEM data sets of moderate or coarse spatial resolutions are freely available, including GTOPO30 (30-arc-second resolution, approximately 1 km), SRTM (3- or 1-arc-second resolution, approximately 90 or 30 m), and ASTER GDEM (1-arc-second resolution, approximately 30 m). In this study, we adopted SRTM of 3-arc-second resolution as the reference DEM in the following experiments.

4 Experimental Results

4.1 Monitoring of Huangtupo Landslide

Huangtupo landslide is a famous ancient giant landslide located in Badong County about 60 km upriver of the Three Gorges Dam. It has not been known as an unstable slope covered with loose debris of about 40–50 m thickness until the geological survey for the resettlement of Badong city in the 1980s. Now, it has been identified as one of the most dangerous large-scale landslides across the Three Gorges Reservoir area, and long-term continuous deformation monitoring is ongoing. In this study, we used archived satellite SAR data sets acquired by C-band ENVISAT/ASAR and L-band ALOS/PALSAR images to retrieve the historic deformations of the Huangtupo landslide. Both two-pass DInSAR and PSI analyses were carried out.

4.1.1 Differential SAR Interferometry (DInSAR) Analyses

The interferometric phase is produced from per-pixel phase difference calculation between two SAR images. It is useful for analyzing wide-area terrain movements along the LOS direction. After removal of topographic phase from the interferograms, fringes should only show up at areas with a high deformation rate. Locations of landslide deformations in large scale can be roughly identified from analysis of fringe density in differential interferograms.

Six pairs of interferometric SAR data covering Badong County were processed to map deformations of Huangtupo landslide over various time spans. Basic information of these data pairs is listed in Table 3. From Fig. 3, we can see that the three differential interferograms with small normal baselines can show deformations at the Huangtupo landslide located on the southern bank of Yangtze River, as outlined by the white squares. Meanwhile, other differential interferograms with long normal baselines are not good enough for identifying landslide deformations. However, DInSAR analysis in the Three Gorges area does not only suffer from strong temporal de-correlations, but also limited by the lack of good reference DEM's, which causes several fringes in the differential interferogram, which could be misinterpreted as surface motions.

4.1.2 Persistent Scatterer SAR Interferometry (PSI)

As mentioned before, three data sets including one ALOS/PALSAR and two ENVISAT/ASAR stacks covering both the old town and the new town of Badong with different system configurations were used to study the Huangtupo landslide in Badong County. Since the capability of penetrating into vegetation of L-band SAR is better than that of C-band systems, CT density of ALOS/PALSAR data set in Badong is higher and more evenly distributed than that of the two ENVISAT/

Table 3 Basic information of InSAR data pairs covering Huangtupo landslide

Pair	SAR sensor	Orbit direction	Date		Normal baseline (m)	Temporal baseline (day)
			Master	Slave		
a	ALOS/ PALSAR	Ascending	2007/ 12/30	2010/ 02/19	156	782
b			2007/ 12/30	2011/ 02/22	1,868	1,150
c	ENVISAT/ ASAR	Ascending	2009/ 05/16	2009/ 11/07	47	175
d			2009/ 05/16	2006/ 07/01	−795	−1,050
e		Descending	2009/ 05/17	2009/ 07/26	−97	70
f			2009/ 05/17	2009/ 11/08	270	175

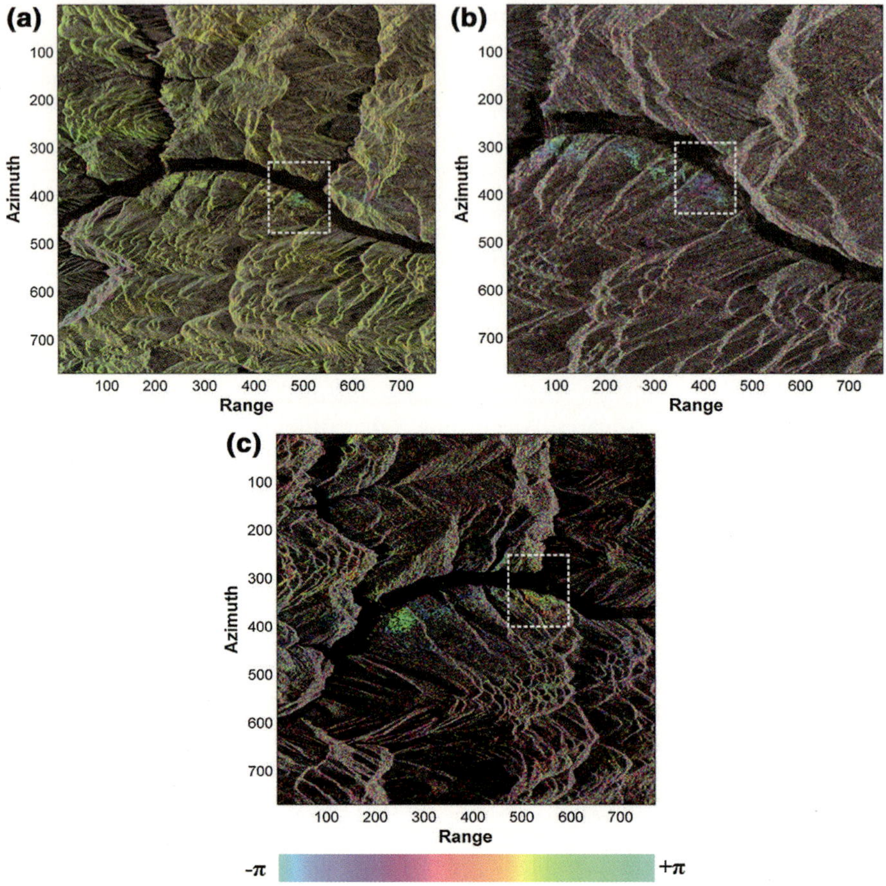

Fig. 3 Differential interferograms after removal of topographic phase, **a**–**c** are interferograms corresponding to data pairs *a*, *c*, and *e* in Table 3, respectively. The *white square* indicates the location of Huangtupo landslide

ASAR data sets, as shown in Fig. 4. Therefore, from a perspective of achieving high density of CT points in rural areas like the Three Gorges, it is more preferable to use L-band SAR data rather than C-band data. As indicated in Fig. 4c, a CT point identified on the county's council building is selected as stable point and all other CT's are calibrated with respect to this reference point.

The most significant deformation observed by ALOS/PALSAR was at Huangtupo landslide in Fig. 4a. The deformation rates derived from the two ENVISAT/ASAR stacks in the LOS direction show generally similar behaviors. The mean velocities from the ascending stack, as shown in Fig. 4b, could be divided into four sub-zones along the Yangtze River, i.e., fast deformation, rather stable, slow deformation, and rather stable from east to west (areas 1–4), respectively. The mean velocities from the descending stack portrayed in Fig. 4c show two fast deformation

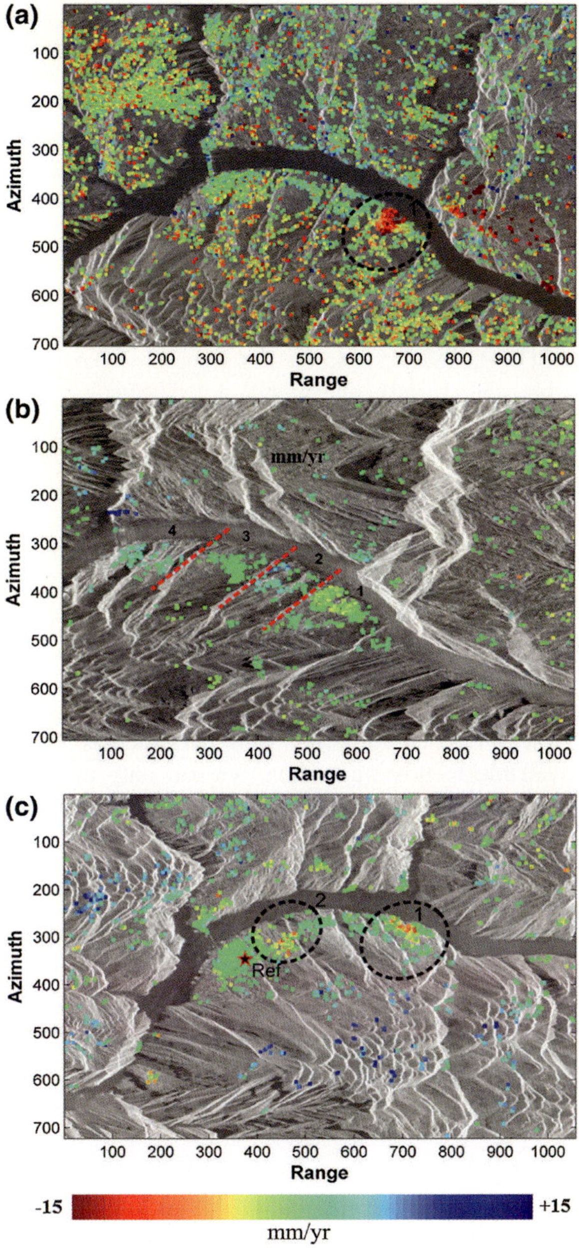

Fig. 4 Deformation velocities at CT points in Badong estimated by PSI analyses: **a** from ALOS/PALSAR data; **b** from ENVISAT/ASAR ascending data; and **c** from ASAR descending data. The *numbered circles* outline the two active landslides. The number *1* in **a** refers to the Huangtupo landslide. The *red lines* in **b** divided the southern riverbank into several significant deformation zones. The *red star* in **c** indicates the location of the reference point

zones (circled areas) located along the southern riverbank of the Yangtze River. From this comparison, we can see that the spatial variability of the mean deformation velocity for the two ASAR stacks is higher than that of the PALSAR stack. This pattern shows the advantage of C-band SAR data over L-band in the sensitivity to deformation measurement due to shorter wavelength when the deformation can be observed in both data stacks.

Apart from these two findings, we can also notice another interesting phenomenon. Along the southern riverbank of the Yangtze River, two areas of significant deformation are identified in Badong area by the two ENVISAT/ASAR stacks acquired from different orbits. PSI results from both ENVISAT/ASAR stacks show the significant deformation in the eastern part of the area (circle 1), but the fast deformation in the western part (circle 2 in Fig. 4c) seems to show up only in the result from the descending stack. According to publicly available documents, the eastern area corresponds to the Huangtupo landslide near the old town, while the western one is located in the new town of Badong. In the result of ascending ALOS/PALSAR stack, the deformation in the eastern site is also clearly identified, while the western one is still not detected. By considering the fact that InSAR analysis can only detect displacements in the LOS direction and taking into account the topographic characteristics of the area around new Badong city, we could infer that it is because the displacement vectors are nearly orthogonal to the LOS vector for the ascending SAR data stacks so that they cannot be detected in the results of ascending stacks. This phenomenon demonstrates that sometimes using just a single InSAR data stack is insufficient to characterize all deformations within a certain region, and it is necessary to carry out joint analyses of multiple stacks gathered from different orbits.

Quantitative accuracy assessment of the mean deformation velocity is essential for evaluating the usability and reliability of CT's in practical applications. However, since no ground truth data such as leveling or GNSS measurements are available in this study area, we can only use an alternative approach, i.e., performing cross-validation among the results from the three InSAR data stacks.

As already mentioned, the deformation velocity measured by the PSI analysis is merely along the LOS direction. Therefore, the results, i.e., the mean deformation velocities at CT points obtained from different data stacks, could not be compared directly. They must be unified into a common reference system before performing the comparison. With a reasonable assumption that the dominant motion vector of a landslide is downward along the slope surface, we chose the direction of slope gradient as the common reference to facilitate interstack comparison. Then, all the LOS velocities were projected onto this direction to obtain downslope velocities using the following equation:

$$V_{\text{slope}} = V_{\text{LOS}} / \cos\theta \tag{20}$$

where V_{LOS} and V_{slope} represent mean deformation velocities along LOS and downslope, respectively, and θ is the intersection angle between the two vectors, i.e., LOS direction and slope surface. After such a conversion, downslope velocities from multiple data stacks could be compared with each other.

As V_{LOS} obtained by PSI is actually a relative measurement with respect to the average deformation velocity across the study area, a calibration operation must be done to offset them to the same reference point before doing the above projection. Location of the reference point is shown in Fig. 4c.

Figure 5 shows the downslope velocities at CT's for the three data stacks. Negative values of V_{slope} stand for a downward motion of the landslide body along the slope surface. It should be noted that the V_{slope} value depends on cos θ. In order to avoid anomalous solutions, a minimum absolute value of cos θ is fixed as 0.3 for this study area. Thus, those CT's with cos $\theta < 0.3$ are discarded in the following assessment of V_{slope} (Herrera et al. 2013). Again, we can see that the active deformation zone of Huangtupo landslide near the old town in the eastern part is

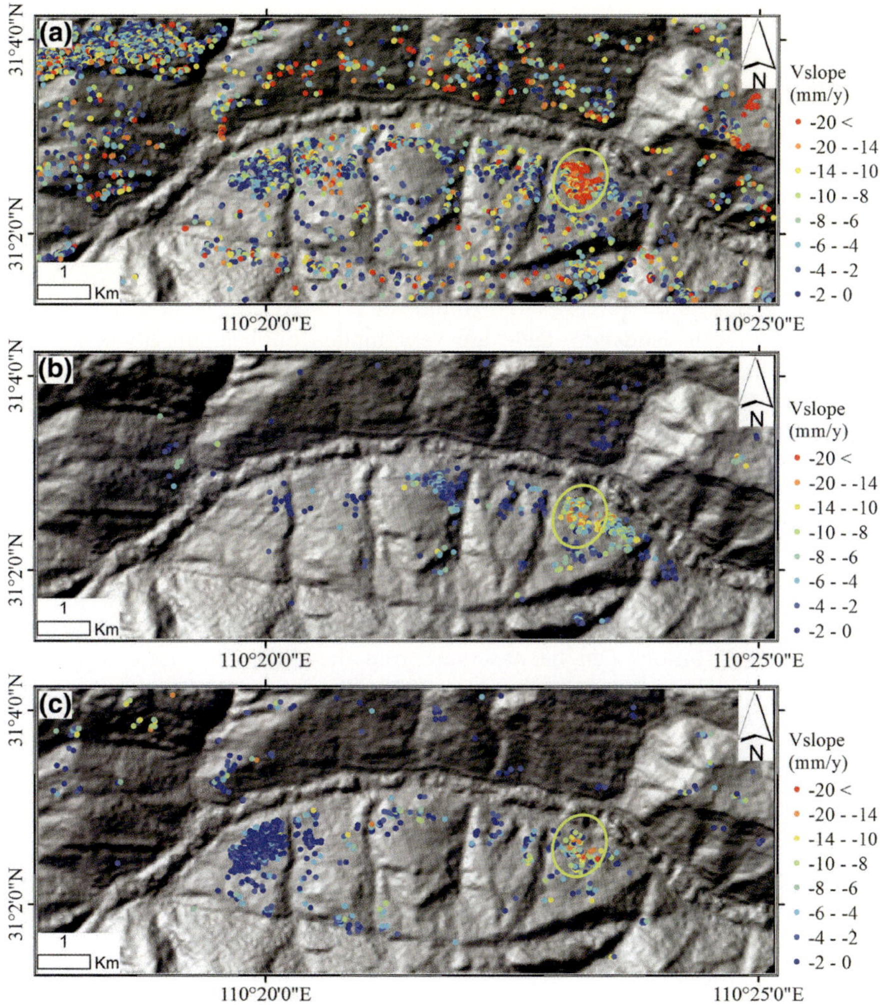

Fig. 5 Downslope mean deformation velocities at CT's derived from the three InSAR stacks. Background image is the hill-shaded map derived from SRTM DEM

clearly identified by all the three stacks, as indicated by the yellow circle in Fig. 5. In order to carry out interstack comparison within this zone, we manually identified nine positions where CT's extracted from the three stacks were spatially close enough to one another so that they can be considered as located over the same ground target. Then, the differences of downslope velocities at each position among the three stacks were obtained and statistics were derived.

The results show that the smallest root-mean-squared difference (RMSD) of 7.0 mm/year is attained to the two ENVISAT/ASAR stacks. A slightly larger RMSD of 8.6 mm/year between the two ascending stacks is achieved, while the largest RMSD of 11.9 mm/year was found to be between the ALOS/PALSAR stack and the ENVISAT/ASAR descending stack. Nevertheless, such differences are still acceptable for landslide deformation monitoring in the Three Gorges region, because in this test area, the quantity of landslide motion is usually at a scale of centimeters per year that is larger than ground subsidence in urban areas.

4.2 Monitoring of Fanjiaping Landslide

Fanjiaping landslide is an ancient landslide located at the southern bank of Yangtze River in Zigui County. It has a large volume of 1.1×10^8 m^3. Although it is a well-known large-scale landslide, there are very limited publications about it, which might be due to the fact that local villagers had been immigrated to other places before the Three Gorges Dam construction, and thus, this landslide has not considered as one of the top-priority slopes to be monitored. Here, we used InSAR method to give some hints about the evolution of Fanjiaping landslide.

High-resolution TerraSAR-X data were used in this analysis. Differential interferograms can be used to identify unstable areas by finding local phase distortions. As it is well known, the short wavelength makes the X-band images prone to suffer from temporal de-correlation. Only interferograms with small temporal and normal baselines were used to visually identify areas of suspect deformations. Very clear phase distortions can be observed from the differential interferograms in Fig. 6, which indicated that Fanjiaping landslide was active during our observation time span. Serious deformation about 1.5 cm occurred on Fanjiaping landslide (see Fig. 6a) from 2nd to 24th January 2009. Tiny phase distortions corresponding to subtle deformations were observed in the two interferograms in Fig. 6b, c.

As we mentioned, temporal de-correlation is always a big concern for InSAR analyses in areas such as Three Gorges covered by dense vegetation. Therefore, all interferograms with high coherence must be fully exploited to ensure the extraction of information as accurate as possible, which makes the SBAS method more suitable for time series analysis in this case. Figure 7 shows the mean deformation rate derived from the TerraSAR-X data sets listed in Sect. 3.2.

Fanjiaping landslide shows a very high deformation rate over 40 mm/year in the upper part. Figure 8 gives a time series plot of displacements at a specific CT extracted from Fanjiaping landslide (marked as *P* in Fig. 7). Deformation about

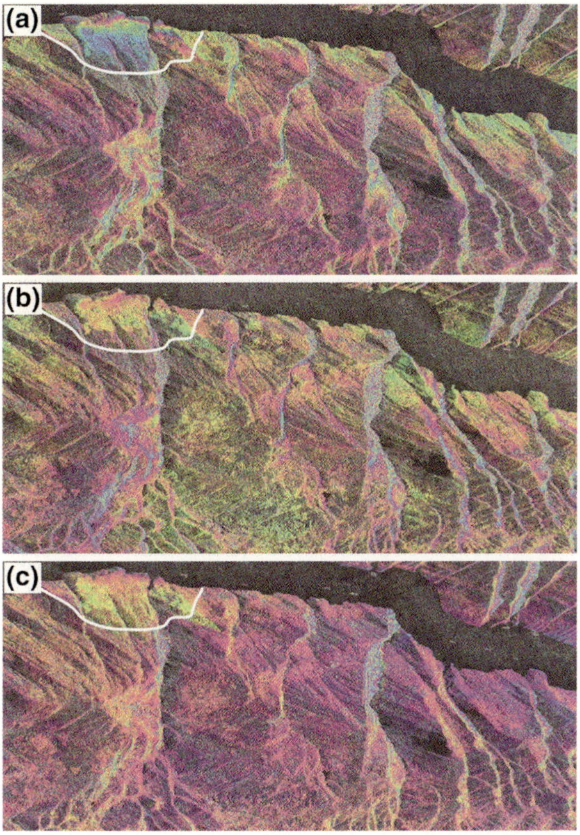

Fig. 6 Differential interferograms of three TerraSAR-X data pairs over different time spans. **a** 2009/01/02 versus 2009/01/24; **b** 2009/11/28 versus 2009/12/09; and **c** 2009/12/09 versus 2009/12/20. The *white curve* indicates the approximate boundary of Fanjiaping landslide

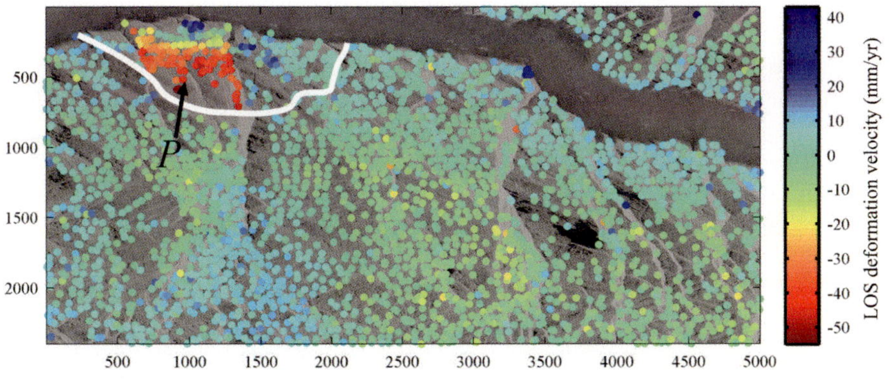

Fig. 7 Mean deformation velocities at CT's identified from TerraSAR-X data sets

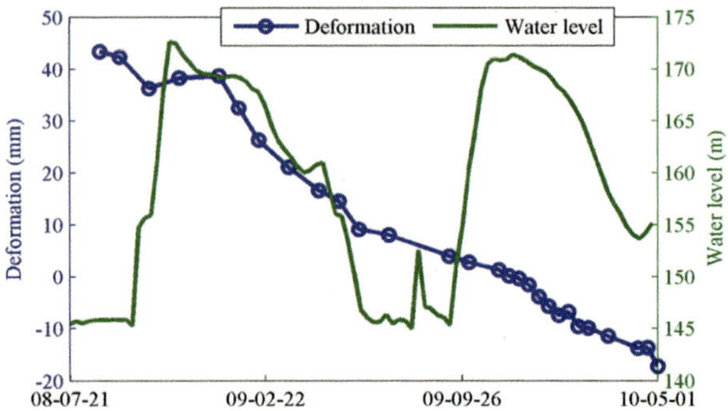

Fig. 8 Time series deformation of a typical CT on Fanjiaping landslide (*in blue*) plotted against the water level of Three Gorges Reservoir

7 cm happened on this point during a time span of almost 2 years. Many studies revealed that there is a strong correlation between the deformation of the landslide and the decline of water level (Liu et al. 2013; Miao et al. 2014). Although there are no obvious signs of strict correlation between deformation and water-level fluctuation, large deformations happened along with the decline of water level from January to May 2009. The relationship between water-level change and landslide displacement will be discussed in the next section on analyses of Shuping landslide.

4.3 Monitoring of Shuping Landslide

The Shuping landslide is located about 47 km west away from the Three Georges Dam. It is also an ancient landslide developed at southern bank of the Yangtze River. It is oriented along the south–north direction with slope ranging from 20° to 30°. The total volume is around 2.6×10^7 m^3 with about 40–70 m thick sliding mass.

DInSAR is firstly employed to visually identify potential active areas. Shuping landslide was judged to be active from phase distortions in the differential interferograms. But in previous study, underestimation was observed in the PSI results on Shuping landslide (Liao et al. 2012), which is mainly due to phase unwrapping errors caused by the large phase gradient. Therefore, InSAR method that is insensitive to displacement in south–north direction is unsuitable for deformation monitoring on Shuping landslide.

Pixel offset tracking method makes use of only amplitude information to measure displacements by subpixel precision image matching, and the step of phase unwrapping is not needed. As the resolution of X-band images is as high as 3 m, the precision can reach about 15 cm if high coherence is maintained. Since 2000, many CR's have been installed on landslides in Three Gorges including Shuping. The backscattering of these corner reflectors (CR's) is very strong and stable over the whole time series and can be used in landslide monitoring. Figure 9a, b shows displacements at CR's along azimuth and range dimensions, respectively, measured by the PTOT method. The eastern part of the Shuping landslide seems very active, where cumulative displacements can reach more than 1 m in azimuth and 0.8 m in range moving toward Yangtze River from 12th August 2008, to 1st May 2010.

Figure 10 shows the time series displacements at corner reflector P1 in Fig. 9. A significant movement on Shuping landslide from April to July 2009 can be easily observed, which showed a high correlation with the decline of water level. This

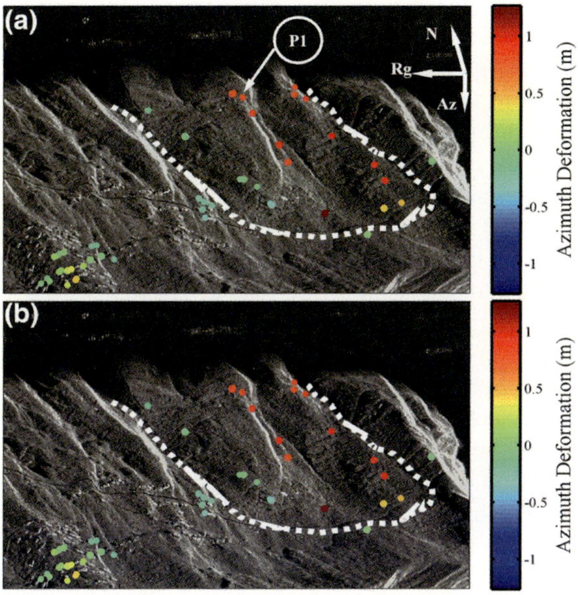

Fig. 9 Cumulative displacements at corner reflectors in **a** azimuth and **b** range measured by PTOT method. P1 is the corner reflector used for the following temporal correlation analysis

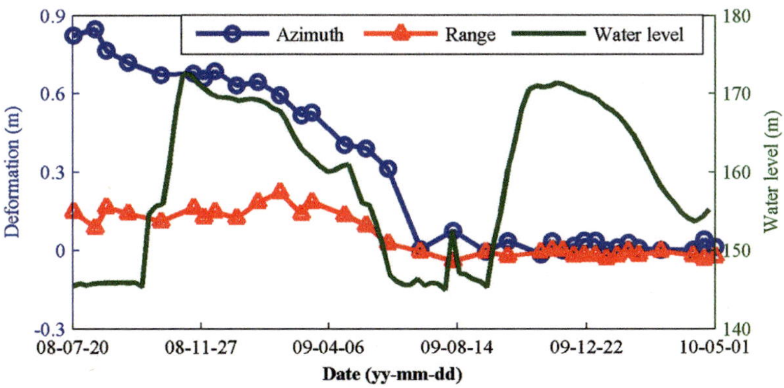

Fig. 10 Temporal cumulative displacements at corner reflector P1 (along range in *red* and along azimuth in *blue*) plotted against reservoir water level (in *green*)

could be explained by a mechanism that causes imbalance between water pressure inside and outside the landslide body when the water level in the reservoir drops down, thus resulting in a significant increase of the slope instability. Therefore, rapid water-level decline might be a key triggering factor of the slope movement.

5 Discussion and Conclusions

Landslides are a common geological problem of great concern in mountainous and hilly regions such as Three Gorges area in China. This makes landslide monitoring a vital task. As more and more SAR satellites were launched into space, large amount of SAR data have become available to scientists. For inaccessible areas, SAR data might be the only useful way for mapping slope displacements to be used for slope failure assessment and, possibly, disaster prediction.

Traditional DInSAR and advanced InSAR methods including PSI and SBAS are based on analysis of the phase information of SAR data. They are useful for monitoring slow movements such as the case of Huangtupo and Fanjiaping landslides in the Three Gorges area.

InSAR methods can measure small deformation with high precision, but problems such as de-correlation and phase unwrapping errors can hinder practical applications of such techniques. For example, the deformation measured by SBAS at Shuping landslide shows obvious underestimation. On the other hand, pixel offset tracking method with relatively lower accuracy can robustly measure displacements in both azimuth and range directions by subpixel-level SAR amplitude image matching. Hence, it is useful for measuring large displacements such as the case of Shuping landslide. A reasonable combination of these methods can provide us more accurate information of the Earth surface displacements.

Various X-band, C-band, and L-band data sets were used to investigate deformations in Three Gorges area. According to our experiments, L-band data less affected by de-correlation are more suitable for InSAR analysis in vegetated areas like the Three Gorges, while high-resolution X-band data are more suitable for pixel offset tracking to measure larger displacements. Since the evolution of landslides is a complex process, SAR images acquired from multifrequency and multiorbits can be combined to inverse 3D displacements, i.e., displacements in northing, easting, and vertical directions.

Unlike other landslides in mountainous areas where deformation is predominantly related to rainfall, water-level fluctuation is considered to be one of the most important triggering factors for those landslides in the Three Gorges area. Significant correlations can be observed between deformation and water-level variations since the huge basin dammed by the Three Gorges Dam has been filled first. Landslide is likely to occur during the period of water-level decline when the water pressure inside the landslide body is higher than that outside. Both Figs. 8 and 10 show a fast-moving trend after the first fulfillment cycle of the basin, where a clear correlation was observed between the displacements and declining of water level. On the other hand, only vague correlation can be observed on Fanjiaping landslide in the second cycle of reservoir operation, which might indicate that the two landslides have different duration times to recover the shear strength. The movement magnitude might also correlate with the speed of water-level decreasing. However, more frequent observations should be collected and analyzed to justify this point.

Acknowledgments This work was financially supported by the National Key Basic Research Program of China (Grant Nos. 2013CB733205 and 2013CB733204), the National Natural Science Foundation of China (Grant Nos. 41271457, 61331016, and 41021061), and the Major Research Program of the Three Gorges Region Geologic Disaster Protection (Grant No. SXKY3-6-4). The authors thank ESA for providing ENVISAT/ASAR data through the Dragon-3 program (id 10569), JAXA for providing ALOS/PALSAR data through ALOS RA3 scientific research projects (PI520 and PI547), and DLR for providing TerraSAR-X data through TerraSAR-X AO project (GEO0606).

References

Angeli, M.-G., Pasuto, A., & Silvano, S. (2000). A critical review of landslide monitoring experiences. *Engineering Geology, 55*, 133–147.

Barazzetti, L., Gianinetto, M., & Scaioni, M. (2014). A new approach to satellite time series co-registration for landslide monitoring. In M. Scaioni (Ed.), *Moderns technologies for landslide investigation and prediction* (pp. 233–249). Berlin: Springer.

Berardino, P., Costantini, M., Franceschetti, G., Iodice, A., Pietranera, L., & Rizzo, V. (2003). Use of differential SAR interferometry in monitoring and modelling large slope instability at Maratea (Basilicata, Italy). *Engineering Geology, 68*, 31–51.

Berardino, P., Fornaro, G., Lanari, R., & Sansosti, E. (2002). A new algorithm for surface deformation monitoring based on small baseline differential SAR interferograms. *IEEE Transactions on Geoscience and Remote Sensing, 40*, 2375–2383.

Cascini, L., Fornaro, G., & Peduto, D. (2009). Analysis at medium scale of low-resolution DInSAR data in slow-moving landslide-affected areas. *ISPRS Journal of Photogrammetry and Remote Sensing, 64*, 598–611.

Cascini, L., Fornaro, G., & Peduto, D. (2010). Advanced low- and full-resolution DInSAR map generation for slow-moving landslide analysis at different scales. *Engineering Geology, 112*, 29–42.

Casu, F., Manconi, A., Pepe, A., & Lanari, R. (2011). Deformation time-series generation in areas characterized by large displacement dynamics: The SAR amplitude pixel-offset SBAS technique. *IEEE Transactions on Geoscience and Remote Sensing, 49*, 2752–2763.

Cojean, R., & Caï, Y. J. (2011). Analysis and modeling of slope stability in the Three-Gorges Dam reservoir (China)—The case of Huangtupo landslide. *Journal of Mountain Science, 8*, 166–175.

Costantino, D., & Angelini, M. G. (2011). Geodetic monitoring applied to a mine area. *Applied Geomatics, 3*, 61–74.

Crosetto, M., Crippa, B., Biescas, E., Monserrat, O., Agudo, M., & Fernández, P. (2005). State-of-the-art of land deformation monitoring using SAR interferometry. *Photogrammetrie Fernerkundung Geoinformation, 6*, 497–510.

Debella-Gilo, M., & Kääb, A. (2012). Measurement of surface displacement and deformation of mass movements using least squares matching of repeat high resolution satellite and aerial images. *Remote Sensing, 4*, 43–67.

Delacourt, C., Allemand, P., Berthier, E., Raucoules, D., Casson, B., Grandjean, P., et al. (2007). Remote-sensing techniques for analysing landslide kinematics: A review. *Bulletin de la Société Géologique de France, 178*, 89–100.

Fastellini, G., Radicioni, F., & Stoppini, A. (2011). The Assisi landslide monitoring: a multi-year activity based on geomatic techniques. *Applied Geomatics, 3*, 91–100.

Ferretti, A., Prati, C., & Rocca, F. (2000). Nonlinear subsidence rate estimation using permanent scatterers in differential SAR interferometry. *IEEE Transactions on Geoscience and Remote Sensing, 38*, 2202–2212.

Ferretti, A., Prati, C., & Rocca, F. (2001). Permanent scatterers in SAR interferometry. *IEEE Transactions on Geoscience and Remote Sensing, 39*, 8–20.

Ferretti, A., Monti-Guarnieri, A., Prati, C., Rocca, F., & Massonet, D. (2007). *InSAR principles: Guideline for SAR interferometry processing and interpretation*. The Netherlands, Noordwijk: ESA Publication. TM-19.

Fornaro, G., Atzori, S., Calo, F., Reale, D., & Salvi, S. (2012). Inversion of wrapped differential interferometric SAR data for fault dislocation modeling. *IEEE Transactions on Geoscience and Remote Sensing, 50*, 2175–2184.

Fourniadis, I. G., Liu, J. G., & Mason, P. J. (2007). Landslide hazard assessment in the Three Gorges area, China, using ASTER imagery: Wushan-Badong. *Geomorphology, 84*, 126–144.

Fruneau, B., Achache, J., & Delacourt, C. (1996). Observation and modelling of the Saint-Étienne-de-Tinée landslide using SAR interferometry. *Tectonophysics, 265*, 181–190.

Grün, A. (2012). Development and status of image matching in photogrammetry. *The Photogrammetric Record, 27*, 36–57.

Handwerger, A. L., Roering, J. J., & Schmidt, D. A. (2013). Controls on the seasonal deformation of slow-moving landslides. *Earth and Planetary Science Letters, 377–378*, 239–247.

Hanssen, R. F. (2001). *Radar interferometry: Data interpretation and error analysis*. Berlin: Springer.

Herrera, G., Gutiérrez, F., García-Davalillo, J. C., Guerrero, J., Notti, D., Galve, J. P., et al. (2013). Multi-sensor advanced DInSAR monitoring of very slow landslides: The Tena Valley case study (Central Spanish Pyrenees). *Remote Sensing of Environment, 128*, 31–43.

Hilley, G. E., Bürgmann, R., Ferretti, A., Novali, F., & Rocca, F. (2004). Dynamics of slow-moving landslides from permanent scatterer analysis. *Science, 304*, 1952–1955.

Hofmann-Wellenhof, B., Lichtenegger, H., & Wasle, E. (2007). *GNSS–global navigation satellite systems: GPS, GLONASS, Galileo, and more*. Berlin: Springer.

Hooper, A., Zebker, H., Segall, P., & Kampes, B. (2004). A new method for measuring deformation on volcanoes and other natural terrains using InSAR persistent scatterers. *Geophysical Research Letters 31*, paper no. L23611.

Hu, X., Wang, T., & Liao, M. (2014). Measuring coseismic displacements with point-like targets offset tracking. *IEEE Geoscience and Remote Sensing Letters, 11*, 283–287.

Jackson, S., & Sleigh, A. (2000). Resettlement for China's Three Gorges Dam: Socio-economic impact and institutional tensions. *Communist and Post-Communist Studies, 33*, 223–241.

Lanari, R., Mora, O., Manunta, M., Mallorqui, J. J., Berardino, P., & Sansosti, E. (2004). A small-baseline approach for investigating deformations on full-resolution differential SAR interferograms. *IEEE Transactions on Geoscience and Remote Sensing, 42*, 1377–1386.

Li, D., Yin, K., & Leo, C. (2010). Analysis of Baishuihe landslide influenced by the effects of reservoir water and rainfall. *Environmental Earth Sciences, 60*, 677–687.

Li, X., Muller, J., Fang, C., & Zhang, Y. (2011). Measuring displacement field from TerraSAR-X amplitude images by subpixel correlation: An application to the landslide in Shuping, Three Gorges Area. *Acta Petrologica Sinica, 27*, 3843–3850.

Liao, M. S., Tang, J., Wang, T., Balz, T., & Zhang, L. (2012). Landslide monitoring with high-resolution SAR data in the Three Gorges region. *Science China Earth Sciences, 55*(4), 590–601.

Liu, J. G., Mason, P. J., Clerici, N., Chen, S., Davis, A., Miao, F., et al. (2004). Landslide hazard assessment in the Three Gorges area of the Yangtze river using ASTER imagery: Zigui-Badong. *Geomorphology, 61*, 171–187.

Liu, P., Li, Z., Hoey, T., Kincal, C., Zhang, J., Zeng, Q., & Muller, J.-P. (2013). Using advanced InSAR time series techniques to monitor landslide movements in Badong of the Three Gorges region, China. *International Journal of Applied Earth Observation and Geoinformation, 21*, 253–264.

Mei, B., Xu, Y., & Zhang, Y. (2013). P- and S-velocity structure beneath the Three Gorges region (central China) from local earthquake tomography. *Geophysical Journal International, 193*, 1035–1049.

Miao, H., Wang, G., Yin, K., Kamai, T., & Li, Y. (2014). Mechanism of the slow-moving landslides in Jurassic red-strata in the Three Gorges Reservoir, China. *Engineering Geology, 171*, 59–69.

Nagler, T., Rott, H., & Kamelger, A. (2002). Analysis of landslides in Alpine areas by means of SAR interferometry. In *Proceedings of IEEE Geoscience and Remote Sensing Symposium, IGARSS '02* (pp 198-200). Toronto, June 24–28, 2002.

Nichol, J. E., Shaker, A., & Wong, M.-S. (2006). Application of high-resolution stereo satellite images to detailed landslide hazard assessment. *Geomorphology, 76*, 68–75.

Peng, L., Niu, R., Huang, B., Wu, X., Zhao, Y., & Ye, R. (2014). Landslide susceptibility mapping based on rough set theory and support vector machines: A case of the Three Gorges area, China. *Geomorphology, 204*, 287–301.

Perissin, D., & Wang, T. (2011). Time-series InSAR applications over urban areas in China. *IEEE Journal of Selected Topics in Applied Earth Observations and Remote Sensing, 4*, 92–100.

Perissin, D., & Wang, T. (2012). Repeat-pass SAR interferometry with partially coherent targets. *IEEE Transactions on Geoscience and Remote Sensing, 50*, 271–280.

Pingue, F., Tammaro, U., Obrizzo, F., & Serio, C. (2013). Vertical ground movements in the Colli Albani area (central Italy) from recent precise levelling. *Applied Geomatics, 5*, 203–214.

Pirotti, F., Guarnieri, A., Masiero, A., Gregoretti, C., Degetto, M., & Vettore, A. (2014). Micro-scale landslide displacements detection using Bayesian methods applied to GNSS data. In M. Scaioni (Ed.), *Modern Technologies for landslide investigation and prediction* (pp. 123–138). Berlin Heidelberg: Springer.

Plateau, T. (2006). Three Gorges Dam: Into the unknown. *Science, 25*, 1034.

Rizo, V., & Tesauro, M. (2000). SAR interferometry and field data of Randazzo landslide (Eastern Sicily, Italy). *Physics and Chemistry of the Earth, Part B: Hydrology, Oceans and Atmosphere, 25*, 771–780.

Raucoules, D., de Michele, M., Malet, J. P., & Ulrich, P. (2013). Time-variable 3D ground displacements from high-resolution synthetic aperture radar (SAR). Application to La Valette landslide (South French Alps). *Remote Sensing of Environment, 139*, 198–204.

Schubert, A., Faes, A., Kääb, A., & Meier, E. (2013). Glacier surface velocity estimation using repeat TerraSAR-X images: Wavelet-vs. correlation-based image matching. *ISPRS Journal of Photogrammetry and Remote Sensing, 82*, 49–62.

Schofield, W., & Breach, M. (2007). *Engineering Surveying* (6th ed.). Oxford: Butterworth-Heinemann.

Shi, X., Liao, M., Wang, T., Zhang, L., Shan, W., & Wang, C. (2014). Express way deformation mapping using high-resolution TerraSAR-X images. *Remote Sensing Letters, 5*, 194–203.

Tantianuparp, P., Shi, X., Zhang, L., Balz, T., & Liao, M. (2013). Characterization of landslide deformations in Three Gorges area using multiple InSAR data stacks. *Remote Sensing, 5*, 2704–2719.

Travelletti, J., Delacourt, C., Allemand, P., Malet, J. P., Schmittbuhl, J., Toussaint, R., & Bastard, M. (2012). Correlation of multi-temporal ground-based optical images for landslide monitoring: Application, potential and limitations. *ISPRS Journal of Photogrammetry and Remote Sensing, 70*, 39–55.

Wasowski, J., & Bovenga, F. (2014). Investigating landslides and unstable slopes with satellite Multi Temporal Interferometry: Current issues and future perspectives. *Engineering Geology, 174*, 103–138.

Wang, F., Zhang, Y., Huo, Z., Peng, X., Araiba, K., & Wang, G. (2008a). Movement of the Shuping landslide in the first four years after the initial impoundment of the Three Gorges Dam Reservoir, China. *Landslides, 5*, 321–329.

Wang, F., Zhang, Y., Huo, Z., Peng, X., Wang, S., & Yamasaki, S. (2008b). Mechanism for the rapid motion of the Qianjiangping landslide during reactivation by the first impoundment of the Three Gorges Dam reservoir, China. *Landslides, 5*, 379–386.

Xia, Y., Kaufmann, H., & Guo, X. (2002). Differential SAR interferometry using corner reflectors. In *Proceedings of IEEE Geoscience and Remote Sensing Symposium*, IGARSS '02 (pp. 1243–1246). Toronto, June 24–28, 2002.

Xia, Y., Kaufmann, H., & Guo, X. (2004). Landslide monitoring in the Three Gorges area using D-InSAR and corner reflectors. *Photogrammetric engineering and remote sensing, 70*, 1167–1172.

Ye, X., Kaufmann, H., & Guo, X. (2004). Landslide monitoring in the Three Gorges area using D-InSAR and corner reflectors. *Photogrammetric Engineering and Remote Sensing, 70*, 1167–1172.

Yin, Y., Wang, H., Gao, Y., & Li, X. (2010). Real-time monitoring and early warning of landslides at relocated Wushan Town, the Three Gorges Reservoir, China. *Landslides, 7*, 339–349.

Radar Technologies for Landslide Detection, Monitoring, Early Warning and Emergency Management

Chiara Del Ventisette, Giovanni Gigli, Veronica Tofani, Ping Lu and Nicola Casagli

Abstract Landslide detection and monitoring represent a starting point to produce hazard and risk maps useful for proper urban planning and emergency management. In this chapter, different applications of radar interferometric techniques are presented to prove their applicability for detection, mapping and monitoring landslides in different geological settings and with different operative conditions. Data acquisition was carried out through satellite (operating in C-band) and ground-based (operating in Ku-band) sensors. In particular, in this chapter, the application of radar interferometry for landslide detection (Arno River basin), for landslide monitoring (Santo Stefano d'Aveto landslide), for landslide hazard scenario definition (Montebeni landslide) and for landslide emergency management (Santa Trada landslide) are presented and discussed. The different applications highlight the capability and suitability of these techniques to work in different operative settings (i.e., different phenomena and geological framework) and for different aims (survey, early warning and emergency assessment).

Keywords Radar interferometry · Ground deformation · Landslide · Rapid mapping · Monitoring · Emergency management

---

C. Del Ventisette (✉) · G. Gigli · V. Tofani · N. Casagli
Department of Earth Science, University of Florence, Via La Pira, 4, 50121 Florence, Italy
e-mail: chiara.delventisette@unifi.it

G. Gigli
e-mail: giovanni.gigli@unifi.it

V. Tofani
e-mail: veronica.tofani@unifi.it

N. Casagli
e-mail: nicola.casagli@unifi.it

P. Lu
College of Surveying and Geo-Informatics, Tongji University, Siping Road 1239, Shanghai 200092, People's Republic of China
e-mail: luping@tongji.edu.cn

M. Scaioni (ed.), *Modern Technologies for Landslide Monitoring and Prediction*, Springer Natural Hazards, DOI 10.1007/978-3-662-45931-7_11

1 Introduction

Ground movements and in particular landslides represent one of the main threats to human life in most mountainous regions in the world.

The occurrence of landslides depends on complex interactions among a large number of partially interrelated factors (geological setting, geomorphologic features, soil properties, land cover characteristics and hydrological impacts). Landslide triggering factors include both natural (intense or prolonged rainfall, earthquakes, volcanic eruptions, rapid snowmelt) and human causes (slope excavation and loading; land use changes, e.g. deforestation; rapid reservoir drawdown; blasting vibrations).

Even though some global landslide databases exist, such as EM-DAT (International Disaster Database; Sapir and Misson 1992), NatCatSERVICE (natural catastrophe loss database; Munich Re; Germany), durham fatal landslide database (DFLD; http://www.landslidecentre.org/database.htm) and many papers have been published on this topic (e.g. Petley et al. 2005; Nadim et al. 2006; Kjekstad and Highland 2008; Kirschbaum et al. 2010), the global impact of slope failures on the population, structures and infrastructures, the economy and the environment remains largely undetermined. In 2004–2010, 2,620 non-seismic, fatal landslides were recorded worldwide, causing a total of 32,322 recorded deaths (more than 4,600 fatalities per year, see Petley 2012) testifying that the existing database underestimate the landslides frequency and the associated fatalities. This suggests that a reassessment of the global landslide risk forecasting and prevention action is needed.

The main step to get this aim is to keep updated landslides inventory at basin/national scale and to monitor known landslides for early-warning purpose.

To measure ground movement displacements, different technologies are available, such as *geotechnical techniques* (extensometer, inclinometers, etc.), *surveying techniques* (manual triangulation, theodolites and GNSS geodetic networks) and *geophysical techniques*. These conventional monitoring techniques (Guzzetti et al. 2012) produce very accurate information but generally limited to a few points on a sliding mass, and they are rarely representative of the whole landslide behaviour.

Furthermore, the two main problems of using conventional techniques, especially for basin-scale landslide detection and monitoring, concern the time and economic cost.

Various *remote-sensing* technologies (passive and active optical sensors, active microwave sensors) can overcome these limitations and, as concerns satellite data, they can provide ground movement information for very large areas (several thousands of square kilometres, see Parker et al. 2011) with temporal and spatial high resolution. In particular radar techniques, both *spaceborne* Differential Interferometric Synthetic Aperture Radar (DInSAR) and Persistent Scatterers Interferometry (PSI) and *ground based* Synthetic Aperture Radar (GBSAR), can operate over wide areas in almost any weather condition, continuously over a long time, providing real-time widespread information with millimetric accuracy without the need of accessing the study area.

This work aims at illustrating, through some selected case studies, the capability of both satellite and ground-based radar techniques to operate in different settings for landslide detection, mapping and monitoring, as well as for early-warning purpose.

2 Radar Techniques

Radar technologies have been developed since the beginning of the twentieth century firstly to remotely detect objects. In the last decades, synthetic aperture radar (SAR) data analysis has been often used for ground movement analysis. The SAR interferometry (Curlander and McDonough 1991) has become, in recent years, one of the emerging techniques for the measurement of ground displacements and an operational tool for monitoring and early warning. SAR techniques are suitable tools in checking/updating inventory maps dealing with mass movement phenomena whose typical velocity values range from few millimetres/year up to 1.6 m/year (i.e. extremely slow landslides according to the classification in Cruden and Varnes 1996). Thanks to more than 20 years covered by radar images, a spatially and temporally detailed characterization of displacement rates is possible today.

The SAR systems used for ground movement detection and monitoring differ both for the platform on which they are installed (spaceborne or ground based) and for the wavelength (frequency) of the emitted signal. Enhancement in wavelength corresponds to a decrease in spatial resolution and measurement precision and an increase in the penetration capacity and stability of the returned signal with respect to the environmental changes.

The radar spaceborne sensors are emitting radio waves in X-band (f = 8–12.5 GHz; λ = 2.4–3.75 cm; i.e. sensor Cosmo-SkyMed by Italian Space Agency and TerraSAR-X by the German Aerospace Center and British National Space Center), C-band (f = 4–8 GHz; λ = 3.75–7.5 cm; i.e. sensor ERS and ENVISAT by European Space Agency and RADARSAT by Canadian Space Agency) and/or L-band (f = 1–2 GHz; λ = 15–30 cm; i.e. sensor JERS and ALOS-PALSAR by Japanese National Space Development Agency).

Although ground-based sensors generally emitting in Ku-band (f = 12.5–18 GHz; λ = 1.6–2.4 cm), some systems operate in band S (f = 2–4 GHz; λ = 7.5–15 cm), C-band (f = 4–8 GHz; λ = 3.75–7.5 cm), band X (f = 8–12.5 GHz; λ = 2.4–3.75 cm) and Ka-band (f = 26.5–40 GHz; λ = 0.75–1.1 cm). The use of Ku-band allows high resolution and high-displacement sensitivity although can cause frequent coherence loss; for this reason, a continuous monitoring is preferred (Monserrat et al. 2014).

DInSAR is a remote-sensing technique that exploits the phase difference (i.e., the interferogram) between SAR data pairs acquired over the same area at different times, allowing the production of deformation maps at centimetre or millimetre accuracy. Power image is related to the amplitude of the backscattered signal. Although the power image is not directly related to the displacement measurement, it allows to understand the main features of the radar image and to find good points for interferometric processing. The *coherence* and hence the quality of the interferometric phase of the observed points is strongly related to the signal-to-noise ratio, i.e. to the amplitude of the backscattered signal. The spaceborne data could be affected by temporal decorrelation due to changes in observed scene (i.e. vegetation, rapid deformation of surface).

A remarkable improvement in the quality of DInSAR results for ground deformation mapping and monitoring is given by the advanced DInSAR techniques (A-DInSAR). These techniques make use of large multi-temporal stacks of spaceborne SAR images acquired over the same area, in order to identify radar targets on which it is possible to detect and measure displacements along the line-of-sight (LOS) direction over time. Two different groups of A-DInSAR techniques can be distinguished on the basis of processing approach: (i) PS-based A-DInSAR techniques (called also PSI) and (ii) Interferogram stacking techniques.

The *PSI techniques* are based on the amplitude and phase information over single isolated reflectors, derived by computing differential interferograms of all the acquisitions with respect to the same reference master image, that are characterized by high temporal signal stability (Wasowski and Bovenga 2014). Different methodologies are based on this approach: Permanent Scatterers InSAR—PSInSAR™ (Ferretti et al. 2000, 2001; Colesanti et al. 2003); SqueeSAR™ (Ferretti et al. 2011); Interferometric Point Target Analysis (IPTA) (Werner et al. 2003; Wegmuller et al. 2004); Persistent Scatterers Pair (PSP) (Costantini et al. 2000, 2009); Stable Point Network (SPN) (Crosetto et al. 2005); Standford Method for Persistent Scatterers (StaMPS) (Hoper et al. 2004, 2007); and QuasiPS (Perissin and Wang 2012).

The *interferogram stacking technique* is based on the information derived from differential interferograms obtained from interferometric pairs with small perpendicular baseline. The decorrelation phenomena that could arise are mitigated by maximizing the number of pixels exploited and improving the signal-to-noise ratio. The small baseline subset (SBAS) (Bernardino et al. 2002; Lanari et al. 2004) is based on this approach.

The scientific literature reports some successful case studies dealing with the detection, mapping and monitoring of landslide phenomena (Hilley et al. 2004; Strozzi et al. 2006; Colesanti and Wasowski 2006; Canuti et al. 2007; Wasowski et al. 2007; Meisina et al. 2007; Cascini et al. 2009; Fornaro et al. 2009; Prati et al. 2010; Herrera et al. 2011; Cigna et al. 2012; Lu et al. 2012; Righini et al. 2012; Tofani et al. 2013; Del Ventisette et al. 2013, 2014; Ciampalini et al. 2014; Wasowski and Bovenga 2014; Zhang et al. 2014).

The ground-based synthetic aperture radar (GBSAR) is a radar-based terrestrial remote sensing imaging system (Tarchi et al. 1997, 2000, 2003; Antonello et al. 2004; Monserrat et al. 2014). GBSAR has proven to be a reliable remote sensing tool which enhances the advantages of the spaceborne InSAR and provides displacement maps with high spatial and temporal resolutions and accuracy, thanks to its high acquisition frequency (up to one image every minute) and a rapid deployment, which makes it a promising tool for managing local (at single slope scale) emergencies. Ground-based radar installations are useful in single hillsides monitoring, while imaging from satellite radar is able to monitor very large areas.

Landslide monitoring applications through GBSAR have been consolidated in the last years (Atzeni et al. 2001; Casagli et al. 2009, 2010; Gigli et al. 2011; Del Ventisette et al. 2012; Intrieri et al. 2013; Nolesini et al. 2013; Bardi et al. 2014; Di Traglia et al. 2014; Mazzanti et al. 2014).

Satellite and ground-based radar interferometry are derived from the same physical principles but in GBSAR, the topographic information does not need to be removed, since the position of the antennas remains the same during different scans (zero baseline condition).

3 Satellite SAR Data for Landslide Detection

Preparation of reliable landslide hazard and risk maps is crucial for hazard mitigation and risk management. In recent years, various approaches have been developed for quantitative assessment of landslide hazard and risk. However, possibly due to the lack of new data, very few of these hazard and risk maps were updated after their first generation. The chosen test site is the Arno River basin, located in central Italy mostly within the Tuscany region. The basin is strongly affected by landslides (Catani et al. 2005; Farina et al. 2006).

From the previously mapped inventory created from 1992 to 2007, it turns out that landslides tend to cluster in the Arno River basin. This clustering effect can also be reflected in the spatial distribution of persistent scatterers which are able to detect slow-moving and extremely slow-moving landslides according to the velocity scale proposed by Cruden and Varnes (1996). In this study, a (semi-)automatic spatial statistical approach, PSI Hotspot and Clustering Analysis (PSI-HCA, Lu et al. 2012), was employed so as to detect spatial clustering of PSI point targets which may be closely associated with landslide occurrences in the mountainous areas of the Arno River basin. PSI point targets processed from four years (from 2003 to 2006) of RADARSAT-1 images based on the PSInSAR™ technique (Ferretti et al. 2001) were deployed for the analysis.

PSI-HCA chiefly consists of two statistical approaches: (1) Getis-Ord G statistics and (2) kernel density estimation. The Getis-Ord G statistics was utilized to estimate the clustering level of PSI point targets while the kernel density estimation was employed to construct a smooth tapered surface of a hotspot. In this study, the scale distance for PSI-HCA was determined by the average value of the shortest distance to a channel and a ridge of each pixel of a 10-m resolution DTM. However, for a more sophisticated estimation, this scale distance is suggested to be evaluated by the bivariate K-function analysis so as to determine the spatial patterns of PSI point targets and landslide occurrences at diverse scales (Lu et al. 2014a).

Figure 1 shows part of the output of PSI-HCA in the Arno River basin. This hotspot map was combined by the outputs from both ascending and descending orbits. The red hotspots indicate the mass movements moving away from the satellite, whereas the blue hotspots reveal the mass movements moving towards the sensor. The deeper the colour is, the higher clustering level of PSI point targets is

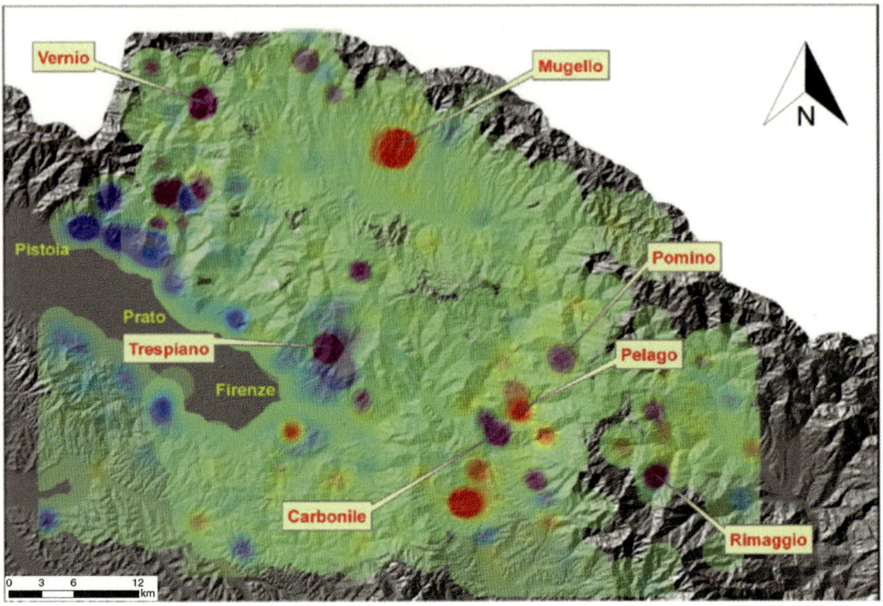

Fig. 1 The output of PSI-HCA covering part of the Arno River basin

detected. The violet hotspots suggest that the mass movements were detected by both ascending and descending orbits, however, with the opposing moving directions. Field surveys were performed to validate most of the labelled landslide-prone regions in Fig. 1 as detected by PSI-HCA.

Hotspot analysis provides an important support for confirming existing landslides in the Arno River basin. The Carbonile landslide (Fig. 1) is a well-known landslide reactivated in 1984. The PSI-HCA approach detected a hotspot in the area and confirmed the previous investigation. At the same time, the hotspot analysis is also an effective tool to detect new extremely slow-moving landslides. In particular, a new landslide was detected in the cemetery of Trespiano (Fig. 1). The hotspot represents a cluster of 678 Radarsat PS's from 2003 to 2006 (Fig. 2). Among them, 211 PS's are from ascending orbits and 467 are from descending orbits. The existing landslide inventory did not report any landslides, and also no evidence of movement is attested by ERS PS's covering the period from 1992 to 2001. The time series of Radarsat PS's suggests that the movement has accelerated since October 2003 (Fig. 2). The velocity of these Radarsat PS's in the cemetery reaches 15 mm/year. Stable and unstable parts can be easily distinguished by checking the velocity distribution of PS's, where a sharp increase of velocity can be seen along the slope (Fig. 2). The landslide is then mapped with the help of the topographic map and aerial photos. The result is validated by field evidence: a field reconnaissance carried out in March 2009 discovered existing damages on walls, roads and structures.

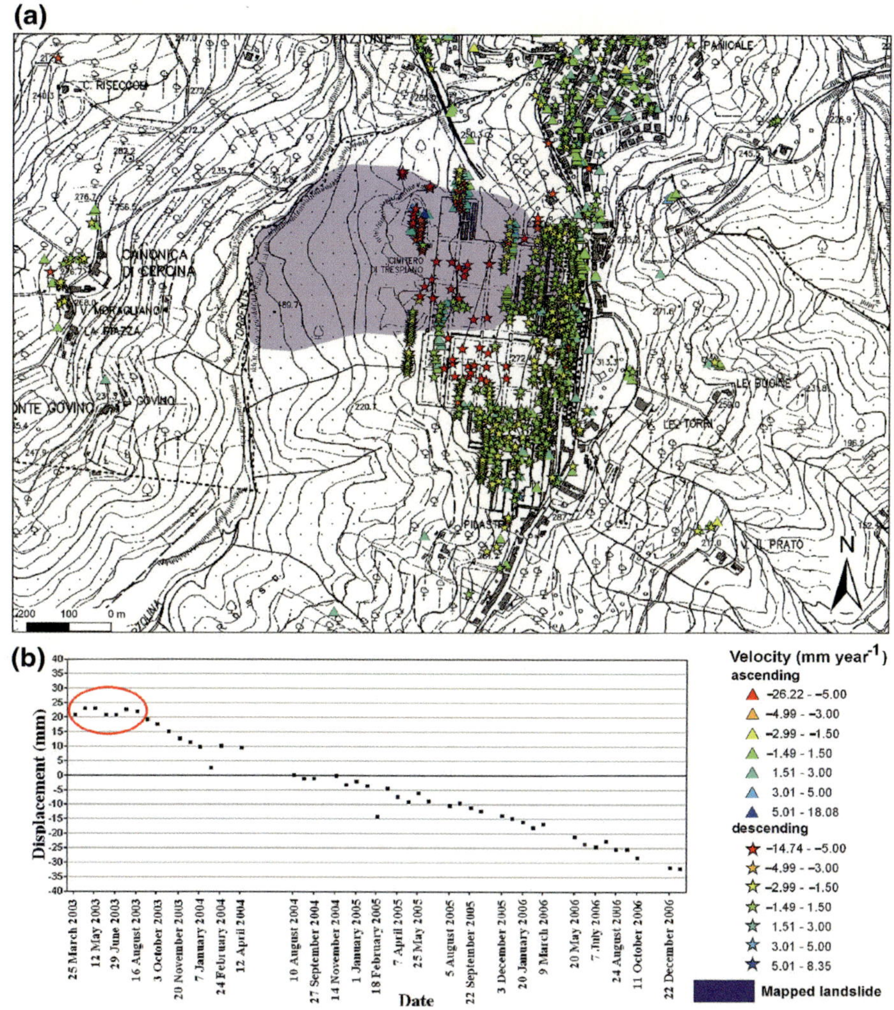

Fig. 2 The Trespiano landslide detected by the PSI-HCA method. **a** Radarsat PS's distribution and mapping results of the new landslide. **b** Time series of PS's located in the northern part of the cemetery

The hotspot detected in the Mugello area (Fig. 1) was attributed to the subsidence due to groundwater pumping and railway tunnel construction instead of the landslide. This may bring one of the major uncertainties for the PSI-HCA approach for landslide detection. This hotspot map was then further used as the input for the landslide hazard and risk assessment in the Arno River basin. In particular, an economic loss of about 3.22 billion Euros throughout the whole basin was predicted for the upcoming 30 years (Lu et al. 2014b).

4 PSI for Landslide Monitoring

An extensive bibliography contains works on the use of PSI for landslide monitoring (Berardino et al. 2003; Singhroy and Molch 2004; Strozzi et al. 2006; Meisina et al. 2007; Fornaro et al. 2009; Prati et al. 2010). In many cases, the PSI data have been integrated with in situ monitoring instrumentation (Farina et al. 2006; Peyret et al. 2008; Liao et al. 2012; Tofani et al. 2013, 2014). The joint use of satellite and ground-based data facilitates the geological interpretation of a landslide and allows a better understanding of landslide geometry and kinematics.

An example of this approach is the Santo Stefano d'Aveto landslide, located in the Northern Apennines (Italy) and affecting the homonym village (Fig. 3). It can be classified as a complex phenomenon that is an earth rotational slide evolving into a flow. The landslide has an extension of 1.3 km^2 and a volume of about 10 million of cubic metres.

From a geological point of view, the main outcropping lithologies are constituted by sandstones with limestones and marls and ophiolitic rocks as basalts and gabbros. Six boreholes have been drilled to investigate the landslide instability and the

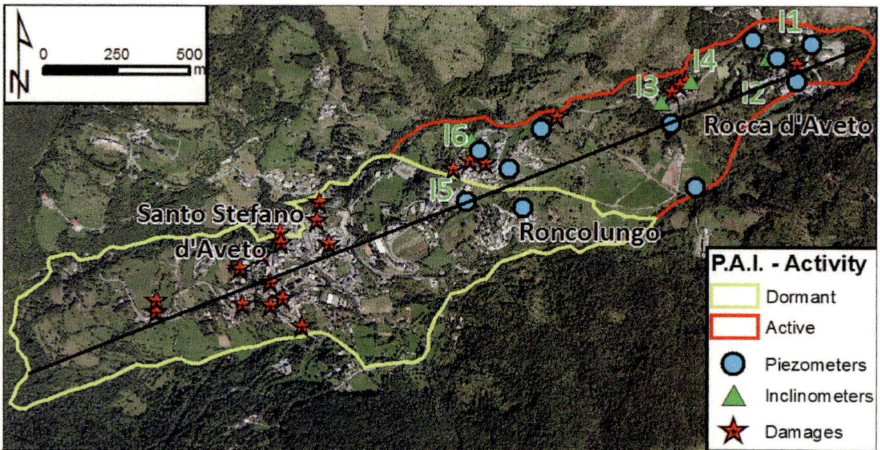

Inclinometers	Mean velocity (mm/yr)	Cumulative displacement (mm)	Depth of the sliding surface (m)
I1	57.8	200	8.5
I2	33.8	35	9
I3	31.2	100	18.5
I4	14.7	30	17
I5	34.1	120	22.5
I6	31.6	30	12.5

Fig. 3 Landslide map with location of the geotechnical monitoring instrumentation and inclinometer measurements (modified from Tofani et al. 2013)

underground geological setting. The results of the boreholes show that the underground geology is characterized by a three-layer stratigraphy, which are starting from the top (Tofani et al. 2013):

(i) coarse debris in a sandy–silty matrix with thin layers of clay material and large blocks of ophiolitic rocks, with a thickness ranging from 25 m to 40 m;
(ii) weathered bedrock with a thickness ranging from 5 m to 20 m; and
(iii) bedrock constituted by the sandstones and rubbles of the Complesso di Monte Veri and slates of the Ottone flysch.

The investigated area is included in the Landslide Inventory Map developed within the PAI (Hydrogeomorphological Setting Plan) of the Po River Basin Authority. The landslide has been classified as active in the upslope portion and dormant in the remaining part where almost all the villages are located (Fig. 3).

The buildings and the main roads of the village are extensively interested by cracks and damages and are often subjected to repairing and consolidation works.

A geotechnical campaign has been carried out from 2000 to 2006, during which six inclinometers and eleven piezometers have been installed inside the landslide perimeter (Fig. 3). The inclinometric measurements have allowed reconstructing the depth of the sliding surface and the rate of movement. The landslide affects the first layer of material composed of colluvium soil made of debris in a sandy–silty matrix. The slip surface is located at about 10 m depth in the upslope portion of the landslide and at about 20 m depth near Santo Stefano d'Aveto village. In general, the depth of the slip surface increases from the upper to the lower portion of the landslide. The annual mean velocities recorded by inclinometers range from 59 mm/year for Sensor I1–15 mm/year for Sensor I4 (Fig. 4). The piezometers have registered a free water table in the debris cover with an average depth of the piezometric surface ranging from a few metres in the upper portion of the landslide up to approx. 20 m at the toe. The measurements present a seasonal variability with a water table lowering during dry seasons and rising in the wet season.

PSI analysis has been carried out making use of ERS-1/2 SAR images spanning from 1992 to 2001 and ENVISAT SAR images spanning from 2002 to 2008, both of them in ascending and descending orbits. The SAR images were processed with the PSInSAR™ approach (Ferretti et al. 2001). The target points within the Santo Stefano landslide have a high density, particularly in correspondence of the urban settlement of the village because the wide availability of bright, stable (i.e. phase coherent) man-made objects (Fig. 4).

The ERS and ENVISAT PS's for both orbits are displayed in Fig. 2. The average absolute velocity of the PS's extracted from ERS descending stack within the landslide is 12.3 mm/year while the maximum value is 37.9 mm/year. The highest velocities, up to 35 mm/year, have been generally measured in the upslope zone of the landslide (Rocca d'Aveto village). In correspondence with the Santo Stefano village, slope velocities vary from a minimum of 6 mm/year to a maximum of 16 mm/year. The average velocity of PS's from ERS ascending data set is approx. 6 mm/year along the satellite LOS, with peaks, recorded nearby the Rocca d'Aveto village, up to 11 mm/year.

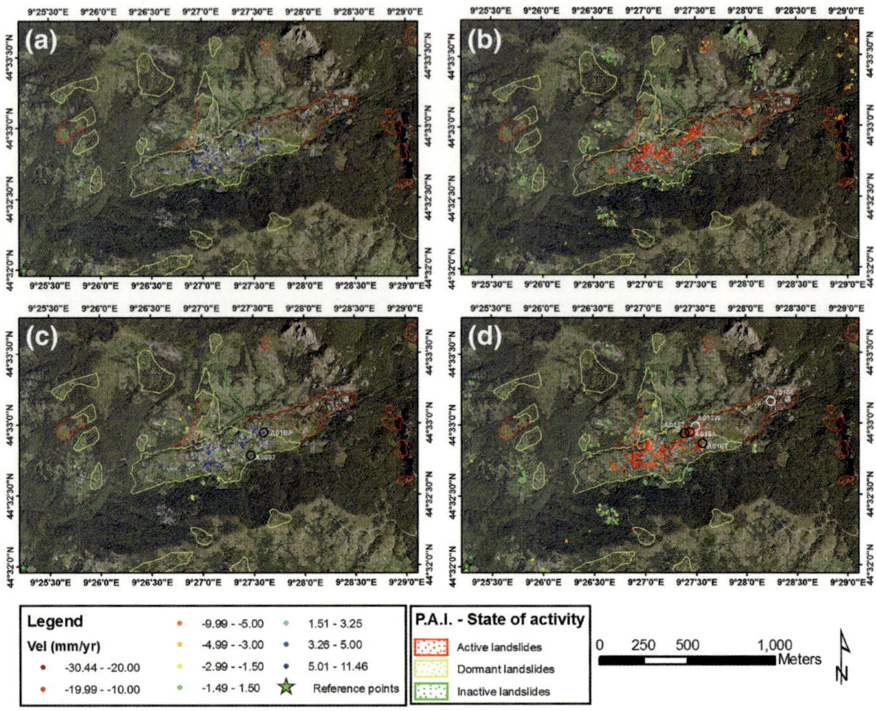

Fig. 4 PSI data available for the Santo Stefano d'Aveto landslide. **a** ERS1/2 ascending (1992–2000). **b** ERS1/2 descending (1992–2001). **c** ENVISAT ascending (2003–2008). **d** ENVISAT descending (2002–2008), (modified from Tofani et al. 2013)

The PS's from ENVISAT descending shows a mean velocity (in absolute value) of 10.2 mm/year with a peak of 19.8 mm/year. The PS's from ENVISAT ascending data set has a mean value of velocity of 3.5 mm/year while the maximum value is around 7 mm/year. Also for ENVISAT data set, deformation rates decrease moving downslope.

The time series of ENVISAT data, spanning the time period from 2002 to 2008, have been compared with the inclinometers' readings, whose measurements covered a period from 2001 to 2006. The Inclinometer I1 has been compared with the PS A0106, whereas the Inclinometer I5 has been compared with the PS A013 W (Fig. 4). In order to perform a representative comparison, the displacement vector of the inclinometers has been projected along the LOS using a simple trigonometry equation:

$$I_{\mathrm{los}} = I_{\mathrm{slope}} \sin\theta$$

where I_{los} is inclinometer displacement along the LOS, I_{slope} is the inclinometer displacement and θ is 23°, the look angle of the SAR sensors. In both cases, the computed velocity values of the inclinometers along the LOS (22 mm/year for I1

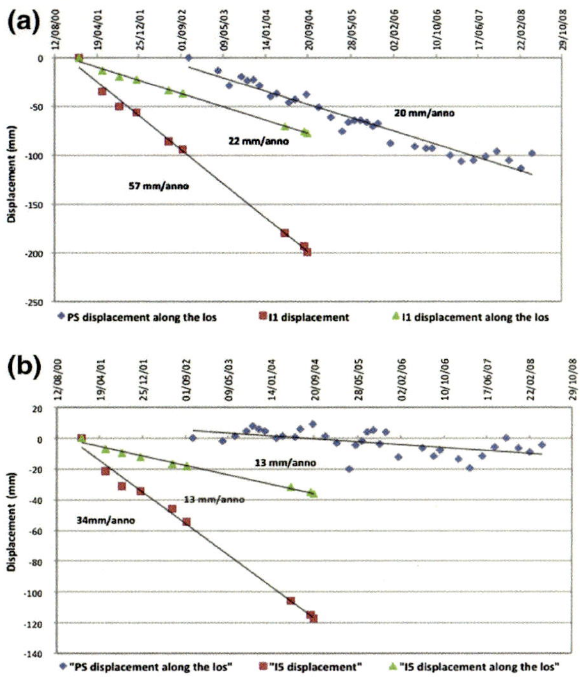

Fig. 5 Comparison of PS's and inclinometer time series of displacement (reproduced with permission from Tofani et al. 2013)

and 13 mm/year for I5) show values quite similar to the mean velocities of closest PS's (20 mm/year and 23 mm/year, see Fig. 5).

The PSI data have been compared also with the piezometer measurements in order to define the relationship among the landslide displacements and the triggering factors. In general, it has been observed that during wet seasons, there is the rise of the water table level in the piezometers and an increase of the displacement rates from PSI technique, while during dry seasons, there is a lowering of the water table level in the piezometers and a decreasing of the displacement rates from PSI technique.

5 GBSAR for Hazard Scenario Definition in Emergency Conditions

Landslides involving large rock masses are always associated with high-risk levels, because their failure may occur catastrophically with long runout. Large rockslides cannot be stabilized at a reasonable cost. Therefore, the most proper measure for risk reduction in settled areas is to set up early-warning systems associated with emergency plans. Correct understanding of a rockslide volume (magnitude) and failure mechanism (with associated energy) is, therefore, very useful, as they influence the type of movement and the runout distance.

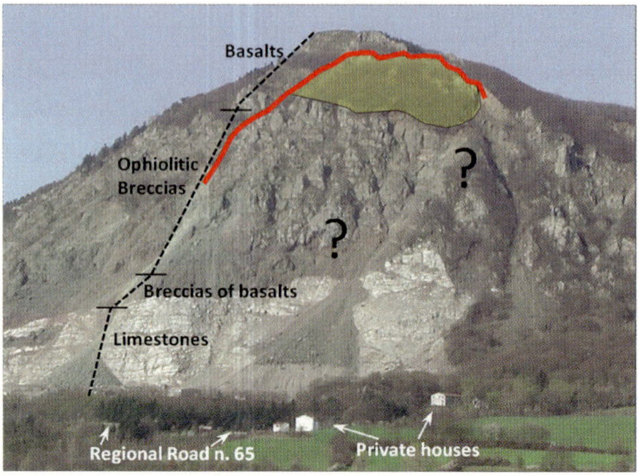

Fig. 6 The eastern slope of Mt. Beni with outcropping materials. *Red line* shows the perimetral crack observed in 2002. The accessible area is shaded in *yellow*

On Mt. Beni (province of Florence, Italy) a reversed ophiolitic sequence outcrops (Fig. 6). The eastern slope had been the object of quarrying activity from the 1940s to the 1980s, when the quarry was abandoned for safety reasons. The total height of the slope, from the quarry floor, is about 340 m.

In April 2002, after a moderate rainstorm, evidence of instability appeared on the slope, with a continuous perimetral crack bounding the upper and right limits of the unstable area, spanning a total length of more than 200 m (red line in Fig. 6).

Due to the lack of evidences and to inaccessibility, the left and lower limits of the rockslide were unknown.

Given the very fast movements, most of the rockslide body was not accessible for the installation of benchmarks or monitoring instrumentation, with the exception of the upper area (yellow sector in Fig. 6), where a conventional distometric monitoring was immediately installed.

In order to extract precious information on the inaccessible sector, the new promising GBSAR technique was employed (Gigli et al. 2011).

From 5th to 19th May 2002, the GBSAR LISA Prototype No. 2 of the Joint Research Centre of the European Commission (JRC) was installed in front of the eastern slope of Mt. Beni at a distance of approx. 2 km. This sensor allowed the investigation of the unstable slope H24, even under bad weather conditions with high rainfall or thick fog.

From the cumulated displacement map (Fig. 7), it was immediately evident that the active area was concentrated in three sectors: (i) the rockslide crown area, corresponding to the perimetral crack; (ii) the lower part of the rockslide, just under the forested area; and (iii) the main debris cone, fed by continuous rock falls from the upper sectors. Thus, the data provided by GBSAR technique allowed the definition, with high precision, of the boundaries of the moving mass and the evolution of the movement in space and time. These outputs were fundamental to calibrate and validate the numerical modelling, which was performed by using a distinct element method for discontinuum modelling (UDEC version 3.1, see Itasca 2000).

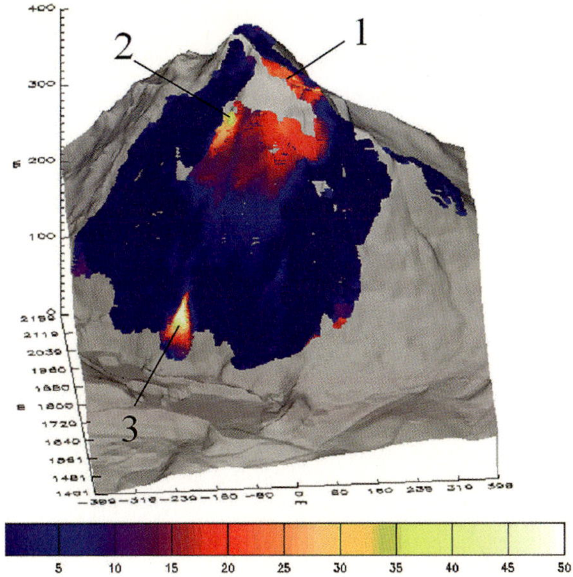

Fig. 7 Cumulated displacement (in mm) map obtained from GBSAR data acquired from 8th to 13th May 2002 (time span: 124 h)

The rock mass characterization and the quantitative description of discontinuities were obtained by geomechanical surveys, according to ISRM suggested methods (ISRM 1978). Discontinuity and rock mass parameters were obtained by using the Hoek and Brown (1997) and Barton and Choubey (1977) criteria, respectively.

The highest vertical displacements were located in the upper and less steep part of the model, were comparatively small in the steepest sector and did not involve the breccias of basalt. As for the horizontal displacements, they did not involve the basaltic breccias and showed a maximum in the median sector of the slope, where the steepness increases. This displacement distribution was in good accordance with the GBSAR monitoring data, as most deformational sectors and activity zones almost completely agreed.

From the numerical model, the upper area was expected to exhibit a prevailing vertical displacement, sinking between sub-vertical cracks, with tension failure prevailing along the steepest lower part experiencing a flexural toppling and associated basal shearing mechanism (Fig. 8).

The complex rock topple/rock slide mechanism resulting from the numerical modelling in agreement with GBSAR data was associated with a paroxistic failure, with a possible long runout distance. Several authors observed that for large rock or debris avalanches, the runout distance increases with landslide volume (Sheidegger 1973; Hsu 1975; Hungr 1990; Nicoletti and Sorriso Valvo 1991; Legros 2002; Morelli et al. 2010). A dynamic analysis was therefore performed using the DAN software (Hungr 1995; Hungr and Evans 1997). The material was assigned a frictional rheologic behaviour, with a dynamic friction angle obtained though the relation proposed by Legros (2002):

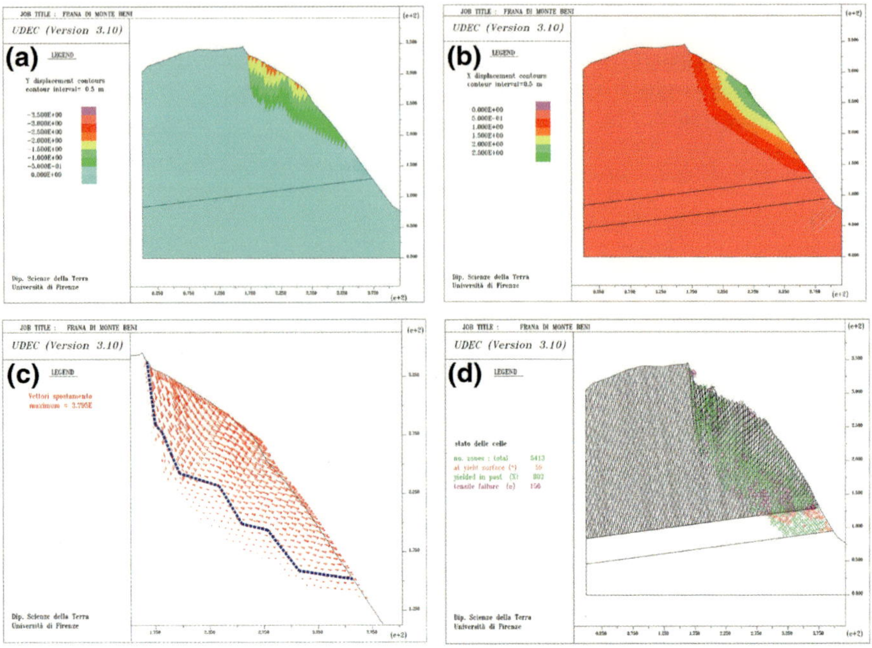

Fig. 8 Distinct element numerical modelling. **a** Vertical displacement contours. **b** Horizontal displacement contours. **c** Displacement vectors. **d** Block model with magnified displacement and zone state (modified from Gigli et al. 2011)

Fig. 9 Runout analysis, considering a rockslide volume of 5×10^5 m^3

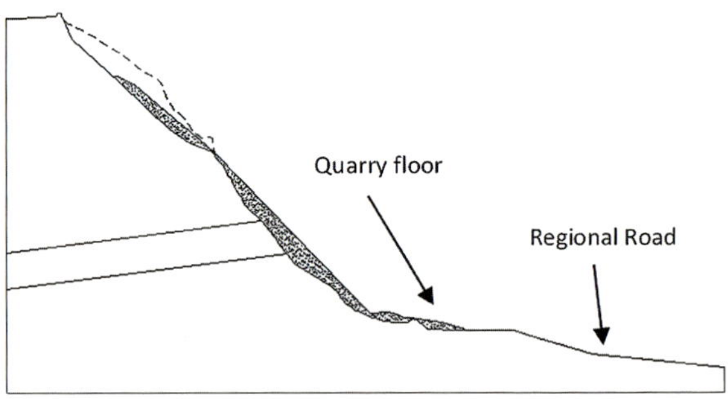

$$H/L = 0.16\,V^{-0.15}$$

where V is the rockslide volume.

Although the planimetric boundary of the rockslide was accurately defined thanks to GBSAR data, the slip surface depth was unknown. Two distinct scenarios were thus analysed, by considering a rockslide volume of 10^6 and 5×10^5 m^3, respectively (Fig. 9).

On 28th December 2002, at 7.30 a.m., the main collapse took place, involving a total volume of about 5×10^5 m^3. The failure mechanism was a rock topple/rockslide type. The detachment area completely corresponded to the one highlighted by the GBSAR observations and the runout was in good agreement with the 5×10^5 m^3 volume scenario.

After the main collapse, the maximum credible scenario was associated with rock falls, due to the presence of large blocks on the slope. Thus, after the construction of a reinforced earth defence barrier, finished on 14th April 2003, the road was opened to traffic again, and the inhabitants were able to return to their houses.

The presented case is one of the first applications of the GBSAR technique for rockslide monitoring, where a complete and successful hazard scenario has been defined and used for civil protection purposes.

6 GBSAR for Early Warning

Thanks to its prompt installation and optimization, one of the main applications of GBSAR systems regards the landslide emergency management for early-warning purpose.

In this section, an application of the GBSAR technique to rapid mapping and to monitor a landslide threatening transportation infrastructures is presented. The application is finalized to define the landslide dynamic and to evaluate possible risk scenario. In the case study presented, the main endangered infrastructure is a motorway, namely the A3 branch, located in Calabria Region (southern Italy). The GBSAR system installed was designed by JRC and successively optimized by Ellegi company.

On 30th January 2009, a mass movement never detected before and located near Santa Trada viaduct (Fig. 10a) caused the closure of that sector of the A3 motorway. From a geological point of view, this area is characterized by Paleozoic micaschists and augen gneiss overlaid by weathered Paleozoic biotite and muscovite granites along with arkosic Pleistocenic sands and conglomerates (Marchetti 1968). The latter are involved in the investigated mass movement. The landslide has a length of 100 m and a maximum width of 90 m. The landslide was triggered by intense rainfall occurred during December 2008 and January 2009 (cumulative monthly rain values of 146.0 mm and 188.6 mm, respectively) that resulted in the saturation of soil for almost the entire period (Del Ventisette et al. 2011). The main aim was to evaluate the possible risk scenarios and to provide early warning. Indeed, despite the landslide did not threaten the roadway directly, its complete collapse would have hit the pillars of the Santa Trada viaduct and, in the event of heavy rainfall, the landslide could be developed in a debris flow affecting the nearby village.

On 31st January 2009, a GBSAR system was installed and after 1 day of signal optimization, the monitoring started. After few days, the GBSAR monitoring allowed to hypothesize two different risk scenarios:

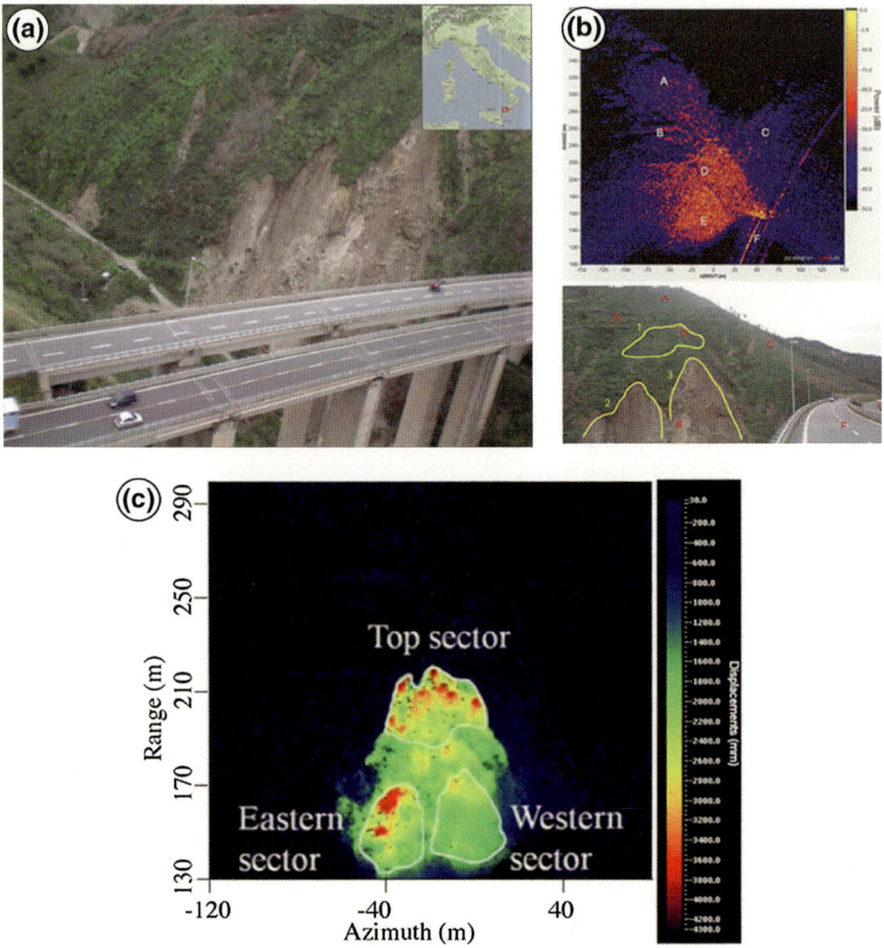

Fig. 10 **a** Picture of Santa Trada landslide nearby the viaduct of the A3 motorway. **b** Radar 'power' image of Santa Trada landslide (*top*) compared to a photograph (*bottom*), both acquired from the GBSAR station. *A* and *B* indicate stable areas; *C* NW slope; *D* and *E* unstable areas; *F* A3 motorway. Landslide sectors: 1—Top Sector; 2—Eastern Sector; 3—Western Sector. **c** Displacement map covering the entire duration of the campaign. Three different sectors are visible. The cumulative displacement recorded during the whole monitoring period of 86 days (from 2nd February 2009 to 29th April 2009) reached the value of 4,320 mm (modified from Del Ventisette et al. 2011)

(i) the landslide would evolve as a series of small collapses due to low-intensity rainfall. Reasonably the collapses would involve three sectors detected by GBSAR. In such a case, collapses would not affect the motorway pillars, even though they could block a minor road located at the foot of the slope; or

(ii) the landslide would fail as a whole. This hypothesis may occur as a result of very intense rainfall. In such a case, the pillars could be damaged producing a serious risk for on way drivers.

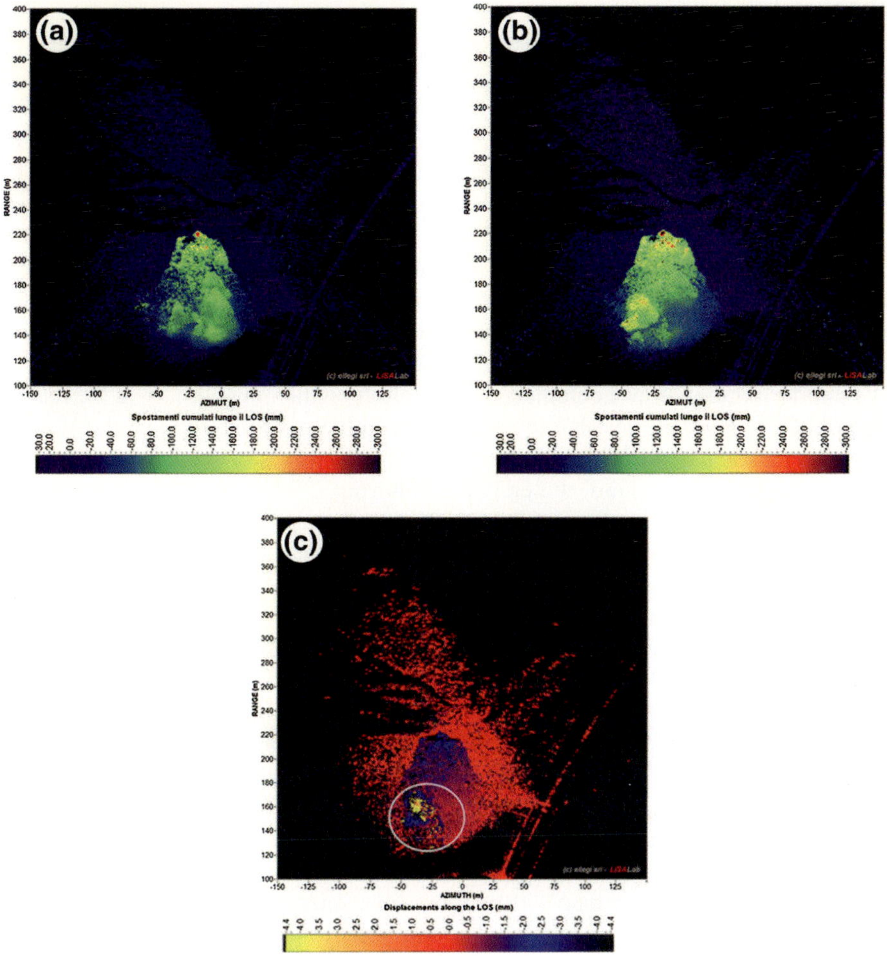

Fig. 11 Example of cumulative deformation map (**a–b**) and interferogram (**c**) achievable from GBSAR data. **a** Cumulative displacement map between 2nd and 3rd February 2009 (time interval 24 h). **b** Cumulative displacement map between 4th and 5th February 2009 (time interval 24 h). **c** Interferogram between 18:08 UT and 18:22 UT of 4th February 2009 (time interval 14 min)

Since 12th February 2009, an internet application with the monitoring result updated every 15 min from GBSAR was working. The continuous monitoring ended on 29th April 2009 when the landslide reached a new equilibrium state.

In order to evaluate the possible risk scenario and to understand the dynamics of the landslide, interferograms (spanning a variable time interval from 7 min to 2 h) and daily displacement maps were elaborated (Fig. 11).

According to the GBSAR first data results, the unstable area has an extension of about 6,000 m^2 (estimated total volume of approx. 18,000–30,000 m^3) and can be divided into 3 sectors (namely Top Sector, Western Sector, and Eastern Sector, see Fig. 10c), characterized by different behaviours in terms of cumulative displacement and dynamics.

During the three months of GBSAR monitoring, the landslide showed several acceleration phases. Between 1st and 5th February, the average velocity of displacements along the LOS of the instrument reached 5.5 mm/h. The landslide continued to accelerate until mid-February up to a value of more than 14 mm/h. On 4th February, a shallow landslide developed in the Eastern Sector. Thanks to the high acquisition frequency of GBSAR, it was possible by means of a mosaic of incremental interferograms spanning a time interval of 2 h, to follow its evolution and to evaluate the deformation velocity (Fig. 11). The superficial collapse reached a maximum velocity of 38 mm/h. This rate was estimated by using interferograms with low time span (7, 14 and 21 min, respectively). By increasing the time span, the interferograms show a complete decorrelation on the 4th February in the Eastern Sector, while it is still possible to estimate the velocity of the Top and the Western Sectors of the landslide.

From mid-February until the end of monitoring (29th April 2009), two periods of heavy acceleration overlapped to a regular decreasing of velocity. During the last month, an average velocity of 0.5 mm/h towards the radar was measured. Before the end of monitoring, the landslide movements were low enough not to represent a risk for the infrastructures in the immediate future.

The rainfall data analysis highlights a close relationship between precipitations and displacements (Del Ventisette et al. 2011).

Field survey has highlighted that also the contiguous slope, facing on the motorway (Sector 'c' in Fig. 10b), was fractured in many points. For this reason, a reprocessing of the GBSAR data was performed to understand and quantify the movements of this sector. Since the alignment between the radar and the assumed direction of movement is not the most favourable, a correction of LOS deformation has to be made (in this case GBSAR recorded 15–35 % of displacement). A cumulative displacement map (spanning a time interval of 59 days between 4th February and 4th April) shows a maximum value of about 5 mm. After the correction, the velocity of deformation results between 8×10^{-3} mm/h and 2×10^{-2} mm/h. Considering the deformation recorded and the thickness of the material involved in this sector (less than 1 m), this slope did not represent a risk for the motorway security.

7 Conclusion

The presented case studies demonstrate the increasing important role played by radar interferometry in landslide management. Indeed, the use of radar interferometry provides to be a very cost and time-efficient technique.

The satellite advanced differential interferometric synthetic aperture radar (A-DInSAR) techniques have demonstrated to be suitable for exploring large land areas and for identifying high-risk zones. It represents an effective tool for remotely measuring and investigating landslide-prone areas, which are characterized by various mechanisms, dynamics and geological/environmental settings.

The PSI technique is suitable for the investigation of extremely slow-moving landslides due to its capability of detecting ground displacements with millimetre precision. However, the interpretation of PSI outputs is sometimes difficult for the large number of possible persistent scatterers (PS's) that may be obtained. The approach proposed in this chapter, the PSI Hotspot and Cluster Analysis (PSI-HCA), is a procedure for mapping landslides efficiently and automatically. PSI-HCA could represent an efficient way to extract useful information from large amount of PSI data especially considering the rapid development of short-wavelength X-band sensors which largely increases PS density and shortens the revisit time.

PSI technique can also have a high potential in the field of landslide monitoring, but the use of this technique alone for this purpose needs to be carefully evaluated. Even if the use of PSI approach may overcome some limitations of conventional monitoring techniques (e.g. large extension of the area to be monitored and high costs, inaccessibility of the area, problems of installation and maintenance), it cannot replace in situ measurements. Based on the results of the application of PSI for landslide monitoring, some general guidelines can be derived. Even though PSI analysis is a valuable technique to monitor landslides, some expedients have to take into account to obtain sound results: (i) correct reconstruction of the actual displacement vector, i.e. the vertical component (VV) and the horizontal E-W component (VE) through the combined use of ascending and descending tracks and (ii) the reconstruction of the link between surface and subsurface displacements combing the PSI results with local underground displacements measures (such as inclinometer readings) in order to compare the PSI-based surface velocities with the actual 3D vectors at different depths in the landslide body.

Ground-based SAR (GBSAR) can provide sequences of deformation maps that allow to study the landslide's temporal evolution, aiming at the detection and definition of kinematic features, even when the acceleration changes are very small, and to evaluate the volume of the material involved in the mass movement. Although the accuracy of the data can slightly decrease, it also granted the possibility of acquiring data during precipitations, which represent the most critical events for slope stability. The accuracy of GBSAR data in mapping and monitoring of landslides is fundamental in risk management and in preparation of emergency plans.

The GBSAR technique is essential for a rapid and detailed definition of risk scenarios in emergency conditions, and, at the same time, it allows the acquisition of all the necessary information without the need of entering dangerous areas, thus ensuring the safety of the involved personnel.

The combined use of satellite and ground-based SAR data may improve the capacity of the monitoring system of slow-moving landslides providing displacement map at different spatial and temporal resolution.

Acknowledgements The research leading to these results received funding from the European Union SAFER and DORIS, ESA TERRAFIRMA Project, and SAR.Net Project founding by National Agency of Civil Protection. We thank the research group of Applied Geology of University of Firenze for their support during projects development. We also wish to thank Ellegi-LiSALab srl for the preprocessing of GB-InSAR data.

References

Antonello, G., Casagli, N., Farina, P., Leva, D., Nico, G., Sieber, A. J., & Tarchi, D. (2004). Ground-based SAR interferometry for monitoring mass movements. *Landslides, 1*, 21–28.

Atzeni, C., Canuti, P., Casagli, N., Leva, D., Luzi, G., Moretti, S., Pieraccini, M., Sieber, A. & Tarchi, D. (2001). Monitoring unstable cultural heritage sites with radar interferometry. In: Sassa K (Ed), UNESCO/IGCP Symposium on Landslide Risk Mitigation and Protection of Cultural and Natural Heritage (p. 257–264), Tokyo, 15–19 January 2001.

Bardi, F., Frodella, W., Ciampalini, A., Bianchini, S., Del Ventisette, C., Gigli, G., et al. (2014). Integration between ground based and satellite SAR data in landslide mapping: The San Fratello case study. *Geomorphology, 223*, 45–60.

Barton, N. R., & Choubey, V. (1977). The shear strength of rock joints in theory and practice. *Rock Mechanics, 10*, 1–54.

Berardino, P., Fornaro, G., Lanari, R., & Sansosti, E. (2002). A new algorithm for surface deformation monitoring based on small baseline differential SAR interferograms. *IEEE Transactions on Geoscience and Remote Sensing, 40*, 2375–2383.

Berardino, P., Costantini, M., Franceschetti, G., Iodice, A., Pietranera, L., & Rizzo, V. (2003). Use of differential SAR interferometry in monitoring and modelling large slope instability at Maratea (Basilicata, Italy). *Engineering Geology, 68*(1–2), 31–51.

Canuti, P., Casagli, N., Catani, F., Falorni, G., & Farina, P. (2007). Integration of remote sensing techniques in different stages of landslide response. In K. Sassa, et al. (Eds.), *Progress in landslide sciences* (pp. 251–259). Berlin: Springer.

Casagli, N., Tibaldi, A., Merri, A., Del Ventisette, C., Apuani, T., Guerri, L., et al. (2009). Deformation of Stromboli Volcano (Italy) during the 2007 crisis by radar interferometry, numerical modeling and field structural data. *Journal of Volcanology and Geothermal Research, 182*, 182–200.

Casagli, N., Catani, F., Del Ventisette, C., & Luzi, G. (2010). Monitoring, prediction and early warning using ground-based radar interferometry. *Landslides, 7*(3), 291–301. doi:10.1007/s10346-010-0215-y.

Cascini, L., Fornaro, G., & Peduto, D. (2009). Analysis at medium scale of low-resolution DInSAR data in slow-moving landslide affected areas. *ISPRS Journal of Photogrammetry and Remote Sensing, 64*, 598–611. doi:10.1016/j.isprsjprs.2009.05.003.

Catani, F., Casagli, N., Ermini, L., Righini, G., & Menduni, G. (2005). Landslide hazard and risk mapping at catchment scale in the Arno River basin. *Landslides, 2*, 329–342.

Ciampalini, A., Bardi, F., Bianchini, S., Frodella, W., Del Ventisette, C., Moretti, S., & Casagli, N. (2014). Analysis of building deformation in landslide area using multisensor PSInSAR™ technique. *International Journal of Applied Earth Observation and Geoinformation, 33*, 166–180.

Cigna, F., Del Ventisette, C., Gigli, G., Menna, F., Agili, F., Liguori, V., & Casagli, N. (2012). Ground instability in the old town of Agrigento (Italy) depicted by on-site investigations and persistent scatterers data. *Natural Hazards Earth System Sciences, 12*, 3589–3603. doi:10.5194/nhess-12-3589-2012.

Colesanti, C., & Wasoswki, J. (2006). Investigating landslides with satellite synthetic aperture radar (SAR) interferometry. *Engineering Geology, 88*, 173–199.

Colesanti, C., Ferretti, A., Prati, C., & Rocca, F. (2003). Monitoring landslides and tectonic motions with the permanent scatterers technique. *Engineering Geology, 68*, 3–14.

Costantini, M., Iodice, A., Magnapane, L. & Pietranera, L. (2000). Monitoring terrain movements by means of sparse SAR differential interferometric measurements. In: Proceedings IGARSS 2000 (Vol. 7, pp. 3225–3227), Honolulu, Hawaii, USA.

Costantini, M., Falco, S., Malvarosa, F., Minati, F. & Trillo, F. (2009) Method of persistent scatterers pairs (PSP) and high resolution SAR interferometry. In: Proceedings IGARSS '09, Cape Town, 12–17 July 2009.

Crosetto, M., Crippa, B., Biescas, E., Monserrat, O., Agudo, M., & Fernández, P. (2005). State-of-the-art of land deformation monitoring using SAR interferometry. *Photogrammetrie Fernerkundung Geoinformation, 6*, 497–510.

Cruden, D. M. & Varnes D. J. (1996) Landslides types and processes. In: A. K., Turner & R. L. Schuster (eds), Landslides: Investigation and mitigation. Transportation research board, National Academy of Sciences (pp. 36–75), Washington DC, Special Report No. 247.

Curlander, J. C., & Mcdonough, R. N. (1991). *Synthetic aperture radar: Systems and signal processing* (p. 647). New York: Wiley.

Del Ventisette, C., Intrieri, E., Luzi, G., Casagli, N., Fanti, R., & Leva, D. (2011). Using ground based radar interferometry during emergency: The case of the A3 motorway (Calabria Region, Italy) threatened by a landslide. *Natural Hazards Earth System Sciences, 11*, 2483–2495.

Del Ventisette, C., Casagli, N., Fortuny-Guasch, J., & Tarchi, D. (2012). Ruinon landslide (Valfurva, Italy) activity in relation to rainfall by means of GBInSAR monitoring. *Landslides, 9*, 497–509.

Del Ventisette, C., Ciampalini, A., Calò, F., Manunta, M., Paglia, L., Reichenbach, P., et al. (2013). Exploitation of large archives of ERS and ENVISAT C-Band SAR data to characterize ground deformation. *Remote Sensing, 5*(8), 3896–3917.

Del Ventisette, C., Righini, G., Moretti, S., & Casagli, N. (2014). Multitemporal landslides inventory map updating using spaceborne SAR analysis. *International Journal of Applied Earth Observation and Geoinformation, 30*, 238–246.

Di Traglia, F., Intrieri, E., Nolesini, T., Bardi, F., Del Ventisette, C., Ferrigno, F., et al. (2014). The ground-based InSAR monitoring system at Stromboli Volcano: Linking changes in displacement rate and intensity of persistent volcanic activity. *Bulletin of volcanology, 76*, 1–18.

Farina, P., Colombo, D., Fumagalli, A., Marks, F., & Moretti, S. (2006). Permanent scatterers for landslide investigations: Outcomes from the ESA-SLAM project. *Engineering Geology, 88*, 200–217.

Ferretti, A., Prati, C., & Rocca, F. (2000). Nonlinear subsidence rate estimation using permanent scatterers in differential SAR interferometry. *IEEE Transactions on Geoscience and Remote Sensing, 38*, 2202–2212.

Ferretti, A., Prati, C., & Rocca, F. (2001). Permanent scatterers in SAR interferometry. *IEEE Transactions on Geoscience and Remote Sensing, 39*, 8–20.

Ferretti, A., Fumagalli, A., Novali, F., Prati, C., Rocca, F., & Rucci, A. (2011). A new algorithm for processing interferometric data-stacks: SqueeSAR™. *IEEE Transactions on Geoscience and Remote Sensing, 49*(9), 3460–3470.

Fornaro, G., Pauciullo, A., & Serafino, F. (2009). Deformation monitoring over large areas with multipass differential SAR Interferometry: A new approach based on the use of spatial differences. *International Journal of Remote Sensing, 30*(6), 1455–1478.

Gigli, G., Fanti, R., Canuti, P., & Casagli, N. (2011). Integration of advanced monitoring and numerical modeling techniques for the complete risk scenario analysis of rockslides: The case of Mt. Beni (Florence Italy). *Engineering Geology, 120*, 48–59.

Guzzetti, F., Mondini, A. C., Cardinali, M., Fiorucci, F., Santangelo, M., & Chang, K. T. (2012). Landslide inventory maps: New tools for and old problem. *Earth-Science Review, 112*, 42–66. doi:10.1016/j.earscirev.2012.02.001.

Herrera, G., Notti, D., Garcia-Davalillo, J. C., Mora, O., Cooksley, G., Sanchez, M., et al. (2011). Analysis with C- and X-band satellite SAR data of the portalet landslide area. *Landslides, 8*, 195–206.

Hilley, G. E., Bürgmann, R., Ferretti, A., Novali, F., & Rocca, F. (2004). Dynamics of slow-moving landslides from permanent scatterer Analysis. *Science, 304*(5679), 1952–1955. doi:10.1126/science.1098821.

Hoek, E., & Brown, E. T. (1997). Practical estimates of rock mass strength. *International Journal of Rock Mechanics and Mining Science & Geomechanics Abstracts, 34*(8), 1165–1186.

Hooper, A., Segall, P. & Zebker, H. (2007). Persistent scatterer interferometric synthetic aperture radar for crustal deformation analysis, with application to Volcan Alcedo, Galapagos. *Journal of Geophysical Research 112*, (paper no. B07407).

Hooper, A., Zebker, H., Segall, P. & Kampes, B. (2004). A new method for measuring deformation on volcanoes and other natural terrains using InSAR persistent scatterers. *Geophysical Research Letters 31* (paper no. L23611).

Hsü, K. (1975). Catastrophic debris stream (sturzstorm) generated by rockfalls. *Bulletin of the Geological Society of America, 86*, 129–140.

Hungr, O. (1990). Mobility of rock avalanches. *Report of the International Research Institute for Earth Science and Disaster Prevention, 46*, 11–19.

Hungr, O. (1995). A model for the runout analysis of rapid flow slides, debris flows, and avalanches. *Canadian Geotechnical Journal, 32*, 610–623.

Hungr, O. & Evans S. G. (1997). A dynamic model for landslides with changing mass. In: Proceedings IAEG International Symposium on Engineering Geology and the Environment (Vol. 1, pp 719–724), Athens, June 1997.

Intrieri, E., Di Traglia, F., Del Ventisette, C., Gigli, G., Mugnai, F., Luzi, G., & Casagli, N. (2013). Flank instability of Stromboli Volcano (Aeolian Islands, Southern Italy): Integration of GB-InSAR and geomorphological observations. *Geomorphology, 201*, 60–69.

ISRM (1978). Commission on the standardization of laboratory and field test—Suggested methods for the quantitative description of discontinuities in rock masses. *International Journal of Rock Mechanics and Mining Science & Geomechanics Abstracts, 15*, 319–368.

Itasca (2000). UDEC: Universal distinct element code—version 3. User's manual. Minneapolis: Itasca Consulting Group Inc.

Kirschbaum, D. B., Adler, R., Hong, Y., Hill, S., & Lerner-Lam, A. L. (2010). A global landslide catalog for hazard applications—Method, results, and limitations. *Natural Hazards, 52*, 561–575. doi:10.1007/s11069-009-9401-4.

Kjekstad, O., & Highland, L. M. (2008). Economic and social impacts of landslides. In D. Sassa & P. Canuti (Eds.), *Landslides—Disaster risk reduction* (pp. 573–587). Berlin: Springer.

Lanari, R., Mora, O., Manunta, M., Mallorqui, J., Berardino, P., & Sansosti, E. (2004). A small baseline approach for investigating deformations on full resolution differential SAR interferograms. *IEEE Transactions on Geoscience and Remote Sensing, 42*, 1377–1386.

Legros, F. (2002). The mobility of long-runout landslides. *Engineering Geology, 63*, 301–331.

Liao, M. S., Tang, J., Wang, T., Balz, T., & Zhang, L. (2012). Landslide monitoring with high-resolution SAR data in the three gorges region. *Science China Earth Sciences, 55*(4), 590–601.

Lu, P., Casagli, N., Catani, F., & Tofani, V. (2012). Persistent scatterers interferometry hotspot and cluster analysis (PSI-HCA) for detection of extremely slow-moving landslides. *International Journal of Remote Sensing, 33*, 466–489.

Lu, P., Bai, S., & Casagli, N. (2014a). Investigating spatial patterns of persistent scatterer interferometry point targets and landslide occurrences in the Arno River Basin. *Remote Sensing, 6*, 6817–6843.

Lu, P., Catani, F., Tofani, V., & Casagli, N. (2014b). Quantitative hazard and risk assessment for slow-moving landslides from persistent scatterer interferometry. *Landslides, 11*, 685–696.

Marchetti, M. P. (1968). *Carta Geologica della Calabria a scala 1*:25000.

Mazzanti, P., Brunetti, A., & Bretschneider, A. (2014). A new approach based on terrestrial remote sensing techniques for rock fall hazard assessment. In M. Scaioni (Ed.), *Modern Technologies for Landslide Investigation and Prediction* (pp. 69–87). Berlin, Heidelberg: Springer.

Meisina, C., Zucca, F., Conconi, F., Verri, F., Fossati, D., Ceriani, M., & Allievi, J. (2007). Use of permanent scatterers technique for large-scale mass movement investigation. *Quaternary International, 171*, 90–107.

Monserrat, O., Crosetto, M., & Luzi, G. (2014). A review of ground-based SAR interferometry for deformation measurement. *ISPRS Journal of Photogrammetry and Remote Sensing, 93*, 40–48.

Morelli, S., Monroi, V. H. G., Gigli, G., Falorni, G., Rocha, E. A., & Casagli, N. (2010). The Tancitaro debris avalanche: Characterization, propagation and modeling. *Journal of Volcanology and Geophysical Research, 193*, 93–105.

Nadim, F., Kjekstad, O., Peduzzi, P., Herold, C., & Jaedicke, C. (2006). Global landslide and avalanche hotspots. *Landslides, 3*, 159–173. doi:10.1007/s10346-006-0036-1.

Nicoletti, P. G., & Sorriso-Valvo, M. (1991). Geomorphic controls of the shape and mobility of rock avalanches. *Bulletin of the Geological Society of America, 103*, 1366–1373.

Nolesini, T., Di Traglia, F., Del Ventisette, C., Moretti, S., & Casagli, N. (2013). Deformations and slope instability on Stromboli Volcano: Integration of GBInSAR data and analogue modeling. *Geomorphology, 180*, 242–254.

Parker, R. N., Densmore, A. L., Rosser, N. J., de Michele, M., Li, Y., Huang, R., et al. (2011). Mass wasting triggered by the 2008 Wenchuan earthquake is greater than orogenic growth. *Nature Geoscience, 4*, 449–452. doi:10.1038/ngeo1154.

Perissin, D., & Wang, T. (2012). Repeat-pass SAR interferometry with partially coherent targets. *IEEE Transactions on Geoscience and Remote Sensing, 50*, 271–280.

Petley, D. N. (2012). Global patterns of loss of life from landslides. *Geology, 40*(10), 927–930. doi:10.1130/G33217.1.

Petley, D. N., Dunning, S. A., & Rosser, N. J. (2005). The analysis of global landslide risk through the creation of a database of worldwide landslide fatalities. In O. Hungr, et al. (Eds.), *Landslide risk management* (pp. 367–374). Amsterdam: A.A. Balkema.

Peyret, M., Djamour, Y., Rizza, M., Ritz, J. F., Hurtrez, J. E., Goudarzi, M. A., et al. (2008). Monitoring of a large slow Kahrod landslide in Alborz mountain range (Iran) by GPS and SAR interferometry. *Engineering Geology, 100*(3–4), 131–141.

Prati, C., Ferretti, A., & Perissin, D. (2010). Recent advances on surface ground deformation measurement by means of repeated space-borne SAR observations. *Journal of Geodynamics, 49*(3–4), 161–170.

Righini, G., Pancioli, V., & Casagli, N. (2012). Updating landslide inventory maps using persistent scatterer interferometry (PSI). *International Journal of Remote Sensing, 33*, 2068–2096.

Sapir, D. G., & Misson, C. (1992). The development of a database on disasters. *Disasters, 16*, 74–80.

Scheidegger, A. E. (1973). On the prediction of the reach and velocity of catastrophic landslides. *Rock Mechanics, 5*, 231–236.

Singhroy, V., & Molch, K. (2004). Characterizing and monitoring rockslides from SAR techniques. *Advances in Space Research, 33*(3), 290–295.

Strozzi, T., Wegmuller, U., Keusen, H. R., Graf, K., & Wiesmann, A. (2006). Analysis of the terrain displacement along a funicular by SAR interferometry. *IEEE Transactions on Geoscience and Remote Sensing, 3*, 15–18.

Tarchi, D., Ohlmer, H., & Sieber, A. J. (1997). Monitoring of structural changes by radar interferometry. *Research in Nondestructive Evaluation, 9*, 213–225.

Tarchi, D., Rudolf, H., Pieraccini, M., & Atzeni, C. (2000). Remote monitoring of buildings using a ground-based SAR. *International Journal of Remote Sensing, 21*(18), 3545–3551.

Tarchi, D., Casagli, N., Fanti, R., Leva, D., Luzi, G., Pasuto, A., et al. (2003). Landslide monitoring by using ground-based SAR interferometry: An example of application to the Tessina landslide in Italy. *Engineering Geology, 68*, 15–30.

Tofani, V., Raspini, F., Catani, F., & Casagli, N. (2013). Persistent scatterer interferometry (PSI) technique for landslide characterization and monitoring. *Remote Sensing, 5*, 1045–1065.

Tofani, V., Del Ventisette, C., Moretti, S., & Casagli, N. (2014). Integration of remote sensing techniques for intensity zonation within a landslide area: A case study in the Northern Apennines, Italy. *Remote Sensing, 6*, 907–924. doi:10.3390/rs6020907.

Wasowski, J., & Bovenga, F. (2014). Investigating landslides and unstable slopes with satellite multi temporal interferometry: Current issues and future perspectives. *Engineering Geology, 174*, 103–138.

Wasowski, J., Casarano, D., Lamanna, C., Bovenga, F., Conte, D., Nutricato, R., Berardino, P., Manzo, M., Pepe, A., Zeni, G., Lanari, R. & Refice A. (2007). A comparative analysis of DIn-SAR results achieved by the SBAS and SPINUA techniques: The Maratea valley case study, Italy. In: Proceedings Envisat Symposium 2007, Montreux, Switzerland, 23–27 April 2007, ESA SP-636.

Wegmuller, U., Werner, C., Strozzi, T., Wiesmann, A. (2004). Monitoring Mining Induced Surface Deformations. In: Proc. IGARSS '04, Anchorage, AK, USA, 20–24 September 2004, Vol. 3, p 1933–1935.

Werner, C., Wegmuller, U., Strozzi, T. & Wiesmann, A. (2003). Interferometric point target analysis for deformation mapping. In: Proceedings IGARSS '03 (pp. 4362–4364), Toulouse, France, 21–25 July 2003.

Zhang, L., Liao, M., Balz, T., Shi, X., & Jiang, Y. (2014). Monitoring landslide activities in the three gorges area with multi-frequency satellite SAR data sets. In M. Scaioni (Ed.), *Modern technologies for landslide investigation and prediction* (pp. 181–208). Berlin, Heidelberg: Springer.

A New Approach to Satellite Time-series Co-registration for Landslide Monitoring

Luigi Barazzetti, Marco Gianinetto and Marco Scaioni

Abstract Image-to-image co-registration is one of the preprocessing steps needed for the analysis of satellite time series. This chapter presents a new approach where all the available images are simultaneously co-registered, overcoming the limits of traditional techniques. This method was tested on the flood and landslide that occurred in Valtellina (northern Italy) during summer of 1987, resulting in the large rockslide of Val Pola. A data set made up of 13 medium-resolution satellite images collected with Landsat-4 and Landsat-5 Thematic Mapper over a period of 30 years was automatically processed. Results showed that the new approach can provide subpixel accuracy close to manual measurements, which today are considered the most accurate method for image registration. The multi-image co-registration method also demonstrated to be atmospheric resistant and robust against land-cover changes, snow, and cloud cover.

Keywords Land-cover changes · Image matching · Least-squares adjustment · Landslide monitoring · Registration

---

L. Barazzetti (✉) · M. Gianinetto · M. Scaioni
Department of Architecture, Built Environment and Construction Engineering, Politecnico di Milano, via Ponzio 31, 20133 Milan, Italy
e-mail: luigi.barazzetti@polimi.it

M. Gianinetto
e-mail: marco.gianinetto@polimi.it

M. Scaioni
College of Surveying and Geo-Informatics, Tongji University, 1239 Siping Road, Shanghai 200092, People's Republic of China
e-mail: marco.scaioni@polimi.it

M. Scaioni (ed.), *Modern Technologies for Landslide Monitoring and Prediction*, Springer Natural Hazards, DOI 10.1007/978-3-662-45931-7_12

1 Introduction

It is well known that satellite remote sensing is a powerful tool to analyze, inspect, and monitor natural hazards (Zlatanova and Li 2008; Fiorucci et al. 2011; Maianti et al. 2014). Satellite images feature several advantages with respect to standard aerial photogrammetric surveys. Indeed, aerial images have to be processed following a time-consuming production workflow: (i) scanning and interior orientation in the case of analogue images (for example, when using archive data); (ii) tie point and ground control point measurement; (iii) bundle adjustment; (iv) 3D stereo plotting or automatic DTM extraction; and (v) orthophoto generation, the latest three representing the main outputs of the photogrammetric process, see Kraus (2008).

Although the recent technologies adopted in aerial photogrammetry (i.e., the use of digital cameras, GNSS/INS positioning systems, and automatic orientation procedures) may overcome some limitations due to sensors, platforms, or costs, such surveys still depend on the organization of the flight mission. This means that when a photogrammetric survey is required over a specific area, the flight plan should be accurately designed and the aircraft should reach the target area from the airport. If the distance to cover is very long or when the environmental or atmospheric conditions prevent to fly, data acquisition cannot be easily carried out.

The recourse to satellite images may avoid such possible concerns, because the sensors continuously operate and no specific missions should be organized. Moreover, the thematic (multispectral) capability of satellite sensors is usually superior to the one of airborne digital cameras, today limited to only four spectral bands in the visible and near-infrared spectrum. Finally, historical satellite data can be included in the analysis, thanks to the availability of large archives of images collected over the past decades (globally not available for aircrafts). For all these reasons, the use of satellite images makes data acquisition simpler, less dependent on logistic constraints, and also cheaper.

Landslide investigation is a typical field where remotely sensed images have become one of the major sources of data. Mantovani et al. (1996) grouped the applications of remote-sensing techniques into (i) *landslide recognition* for the production of inventory maps and databases; (ii) *landslide monitoring*; and (iii) *spatial analysis and hazard zonation*. In particular, landslide monitoring requires the analysis of multiple data gathered over a period of many years. This task can be defined as the comparison of different conditions (extension, velocity, topography, soil humidity, and other parameters) in order to assess the landslide activity, see Metternicht et al. (2005). Among these indicators, the measurement of surface deformation has a primary relevance, requiring suitable image resolution and timely acquisition depending on the velocity of the geoprocess under investigation (Khairunizza-Bejo and Petrou 2010; Scaioni et al. 2014b).

As reviewed in Scaioni et al. (2014a), many studies have been carried out on the use of new satellite data for landslide investigations. However, few efforts have been made on the analysis of the quality aspects of such applications. When using geospatial data, the accurate registration of all the data sets into the same reference systems is a fundamental prerequisite for any monitoring purposes, especially when time series are used (Gosthtasby 2005; Gianinetto 2012), since image registration issues could influence the final quality of the analysis. The same problem can be faced in the production of multi-temporal (Nayak and Zlatanova 2008) and seasonal landslide inventory maps, see Guzzetti et al. (2012).

According to Brown (1992), *image registration* can be defined as 'the process of overlaying images (two or more) of the same scene taken at different times, from different viewpoints, and/or by different sensors.' As introduced, image registration has an essential role in remote sensing when multi-date images are used, see Le Moigne et al. (2011). This phase is usually carried out manually when a few images are involved; otherwise, automatic data processing should be preferred in the case of large data sets. *Manual registration* methods are based on the interactive selection of a set of control points (CP's) between a reference image (or 'master' image) and several other images (or 'slave' images). This operation is not trivial, and expert operators measure precise homologous points with a uniform geometric distribution. Obviously, if a huge amount of images need to be co-registered, this task may become heavy and very time-consuming. In such case, *image-to-image* co-registration algorithms are used for the automatic detection of corresponding CP's. Traditionally, pairs of images (the master and the slave) are co-registered at the same time (Smith and Atkinson 2001). Consequently, if the data set is composed of several images, any of them has to be independently co-registered with respect to the master.

All *automatic registration* algorithms should be able to manage images collected under different illumination, in different years or seasons, with different viewing angles, with uneven shadows, as well as cloud and snow cover. In addition, the registration algorithm should be robust against the presence of small changes in the scene, like the ones due to moving objects (e.g., vehicles, temporary objects, etc.), vegetation changes, or new/modified constructions. In some cases, this step is a kind of involved task and manual measurements become mandatory. The extracted CP's are then used to estimate a mapping function between the different images. Overall, the problem consists in (i) the preliminary selection of an appropriate mathematical model for image registration and (ii) the estimation of its unknown parameters. A typical workflow includes the extraction of elements like corners, edges, line intersections, or centroids of specific regions. These features should be visible in two or more images of the series, meaning that stability in time and invariance to geometric and radiometric changes are needed. The number of features (and their distribution) has to be sufficient to warrant a reliable estimation of the unknown transformation parameters. *Image matching* (see Scaioni et al. 2014b) is then carried out with different methods that exploit the intensity value of adjacent pixels, their spatial pattern, or description by means of mathematical vectors. The scientific literature is vast (a review is reported in Gosthtasby 2005; Le Moigne et al. 2011; Grün 2012), and some of the most popular techniques are *correlation-like* methods (Pratt 1991), *mutual information* (Pluim et al. 2001), *Fourier transform methods* (Castro and Morandi 1987), *relaxation methods* (Price 1985), and *spatial relations* (Goshtasby et al. 1986).

Once a set of CP's is detected, a mapping model can be estimated. Obviously, automated methods may produce gross errors (outliers). Thus, robust estimation techniques are recommended. The final mapping function is usually assumed as valid for the whole image in order to estimate a global transformation via Least-squares techniques, see Mikhail et al. (2001). In the case of medium-resolution satellite imagery, low-degree polynomial models are mostly used, including similarity, affine, homography, and second-order or third-order polynomials. Rational function models (RFM) are instead preferred in the case of high- and very high-resolution imagery (Dowman and Tao 2002; Gianinetto and Scaioni 2008; Poli and Toutin 2012) that still needs terrain correction.

A new paradigm to automatic image-to-image registration involves the concept of global adjustment, where all the satellite images are simultaneously processed, thereby avoiding the single matching between one master and the slaves. This approach is different from all the other existing methods because it provides a global mapping function for the whole data set and has the advantage of a higher data redundancy, which helps the detection of blunders and increases the reliability of results. In addition, slave images that cannot be directly registered to the master, because no correspondences could be detected, can be processed as well. Alike the traditional techniques, one image has to be set as reference to establish the datum to which all other images will be co-registered. More details can be found in Barazzetti et al. (2014a, b).

2 Algorithm Description

As mentioned, the method matches features among all the images of a time series. In a first step, the matching is carried out between image pairs and all the image combinations are analyzed. CPU time depends on N^2 (where N is the number of images) although the process can be parallelized as the matching of a specific image pair does not influence the others. In the case study presented later on, thirteen images provided $N \times (N - 1)/2 = 78$ combinations.

The matching phase has to take into consideration the following matters:

(i) Images can be collected in different seasons and years, with different cloud fraction, snow cover, or with significant land-cover changes. Furthermore, the images may be recorded with different sensors. Thus, the algorithm should be robust against radiometric and geometric changes;

(ii) Corresponding features among different images have to be extracted automatically, and consequently, the algorithm must be able to remove gross errors without any user's interaction;

(iii) Multiple satellite images are processed at the same time; thus, the algorithm must be able to match the same features in several images (multi-image matching);

(iv) Images have to be processed at their original resolution and extension, without any spatial or spectral down-sampling; and
(v) Subpixel precision is required.

As the method described in this chapter was developed for fully automatic image processing, any preliminary data processing was avoided (e.g., no atmospheric correction was applied).

2.1 Detection of Corresponding Image Features

The workflow starts with the extraction of image features with the Speeded up Robust Features (SURF, see Bay et al. 2008) operator. This algorithm implements two main functions: a *detector* capable of finding interest features which are well characterized with respect to the surrounding background and a *descriptor* which associates a vector of information to any detected features. Descriptor vectors can be used for matching features between different images without any preliminary information such as seed points or other manual measurements, see Teke and Temizel (2010) and Bouchiha and Besbes (2013). This kind of image matching procedure can be classified as *feature-based matching* (FBM), see Apollonio et al. (2014).

Given a generic image pair (i, j) in which m and n features were detected with descriptors $\boldsymbol{D}_m$ and $\boldsymbol{D}_n$, the matching procedure compares all the descriptors of the image i with all the ones of the image j by using the Euclidean distance $\|\boldsymbol{D}_m - \boldsymbol{D}_n\|$. The couple of features with the minimum distance are selected as corresponding features. On the other hand, owing to the presence of repetitive patterns in the images, one feature on the master image may be closely similar to more than one feature on the slave. To limit the risk of false matches, a 'ratio test' is included, following Barazzetti et al. (2010). A constraint between the first best and the second best candidates is considered to avoid the selection of ambiguous corresponding features:

$$\|\boldsymbol{D}_m - \boldsymbol{D}_n\|^{\text{first}} < 0.75 \cdot \|\boldsymbol{D}_m - \boldsymbol{D}_n\|^{\text{second}} \quad (1)$$

This strategy does not completely avoid the presence of outliers in the set of extracted corresponding features, leading to a weak computation of the mapping transformation. The solution needs to be sought with robust techniques for outliers rejection able to handle a large percentage of gross errors. The method implemented is based on the popular high-breakdown point estimator RANSAC, see Fischler and Bolles (1981).

In the case of medium-resolution satellite images, a 2D similarity transformation incorporating two shifts, a scaling and a rotation in the plane, can been used. This is motivated by the geometrical effect given by this transformation, where the shape (angles and curvatures) is preserved. Indeed, orthorectified (terrain-geocoded)

Landsat images could be potentially affected by the following sources of errors: (i) an overall translation (e.g., when corner coordinates generally do not match exactly); (ii) row and column counts will usually be different; and (iii) translational differences will exist with respect to the upper left corner of the image (Wahed et al. 2013).

The automatic strategy here described retrieves a large number of image correspondences (usually hundreds of features) that are more than enough for the estimation of the transformations commonly used for satellite images. This data redundancy also helps the outlier rejection stage.

2.2 Estimation of the Geometric Model

The FBM operated by means of SURF is exhaustively applied to all the image pair combinations. Consequently, pairwise corresponding features are found. On the other hand, some of these features may be shared between more than two images. Thus, such 'multiple' features are recognized and reordered by comparing the image coordinates in a second stage. The availability of multiple features is extremely important to increase the local reliability of the observations (Mikhail et al. 2001; Kraus 1997), and consequently, it helps the localization of residual outliers. Indeed, RANSAC is quite efficient to cope with a large fraction of even large outliers after the pairwise FBM, but it is not optimal to remove small errors which could degrade the quality of the transformation estimate. This task is carried out during a simultaneous adjustment including all extracted observations (i.e., feature coordinate measurements) after feature reordering. At this stage, all the 2D similarity transformations used to map all the slave images onto the master are contemporarily estimated.

The *functional model* for image registration is based on the following *observation equations* describing a 2D similarity transformation:

$$\begin{aligned} x'_{ik} &= s_j x_{ij} \cos\alpha_j - s_j y_{ij} \sin\alpha_j + t_{xj} = a_j x_{ij} - b_j y_{ij} + t_{xj} \\ y'_{ik} &= s_j x_{ij} \sin\alpha_j + s_j y_{ij} \cos\alpha_j + t_{yj} = b_j x_{ij} + a_j y_{ij} + t_{yj} \end{aligned} \tag{2}$$

where i is the feature index, j and k the image indexes. Equation (2) expresses that each slave image is shifted (t_{xj}, t_{yj}), rotated (α_j), and scaled (s_j) with respect to the master image. Such equations are nonlinear, but the simple substitution with coefficients a_j and b_j provides a system of linear equations.

Two types of observation equations can be formulated depending on the nature of the considered feature. If the feature is shared between a generic slave image j and the master image R, the unknowns (underlined in the following formulas) are only the four transformation parameters of the similarity transformation mapping image j to image R:

$$\begin{aligned} x'_{iR} &= \underline{a}_j x_{ij} - \underline{b}_j y_{ij} + \underline{t}_{xj} \\ y'_{iR} &= \underline{b}_j x_{ij} + \underline{a}_j y_{ij} + \underline{t}_{yj} \end{aligned} \tag{3}$$

If corresponding features are found between two generic slave images, they become a sort of *tie point* (using the photogrammetric terminology, see Kraus 2008) which gives rise to a couple of equations where feature coordinates projected onto the reference image are estimated (unknowns are underlined):

$$\begin{aligned} \underline{a}_j x_{ij} - \underline{b}_j y_{ij} + \underline{t}_{xj} - \underline{x}'_{iR} &= 0 \\ \underline{b}_j x_{ij} + \underline{a}_j y_{ij} + \underline{t}_{yj} - \underline{y}'_{iR} &= 0 \end{aligned} \tag{4}$$

The solution here proposed is the same obtainable with a weighted adjustment based on *pseudo-observations*, assuming that features on the master image remain fixed ($\sigma = \pm 0$ pixel) and those of slave-to-slave connections have the same precision (for more details, see Barazzetti et al. 2014a). Equations can be cast into in the following matrix form:

$$\mathrm{v} = \mathbf{A}\begin{pmatrix} \mathrm{x}_1 \\ \mathrm{x}_2 \end{pmatrix} - \begin{pmatrix} y_1 \\ 0 \end{pmatrix} = \mathbf{A}\mathrm{x} - \mathrm{y} \tag{5}$$

where the vector v contains the residuals of the observations; the design matrix **A** groups all the coefficients of the observation equations; vector x_1 contains the ordered unknown transformation parameters and x_2 the image coordinates of the tie points reprojected onto the master image, and vector y contains the coordinates of features measured directly on the master in the case of Eq. (3). The normal matrix can be defined as follows:

$$\mathbf{N} = \mathbf{A}^{\mathbf{T}}\mathbf{A} = \begin{bmatrix} \mathbf{A}_1 & \overline{\mathbf{N}} \\ \overline{\mathbf{N}}^{\mathbf{T}} & \mathbf{A}_2 \end{bmatrix} \tag{6}$$

N has a particular banded form, where $\mathbf{A_1}$ is a hyper-diagonal matrix with sub-matrices 4 × 4, whereas $\mathbf{A_2}$ is a diagonal matrix (note that **N** has zero elements). $\overline{\mathbf{N}}$ contains the off-diagonal elements. Thanks to the normal matrix, the normalized system of equations can be derived in the forms:

$$\mathbf{N}\mathrm{x} = \mathbf{A}^{\mathbf{T}}\mathrm{y} \tag{7}$$

where $x = [\mathrm{x}_1\ \mathrm{x}_2]^{\mathbf{T}}$. The Least-squares solution is computed as follows:

$$\hat{\mathrm{x}} = \mathbf{N}^{-1}\mathbf{A}^{\mathbf{T}}\mathrm{y} \tag{8}$$

The solution allows the estimate of the residuals in Eq. (5), which are due to the fact that the observed image coordinates cannot perfectly fit the estimated model due to measurement errors:

$$\widehat{\mathrm{v}} = \mathbf{A}\widehat{\mathrm{x}} - \begin{pmatrix} \mathrm{y}_1 \\ 0 \end{pmatrix} \tag{9}$$

The analysis of residuals can be exploited to detect outliers by means of standard *data snooping* technique, see Kraus (1997).

The estimated sigma naught is given by

$$\sigma_0 = \sqrt{\frac{\mathrm{v}^{\mathrm{T}}\mathrm{v}}{r}} \tag{10}$$

where r is the global redundancy, which is the difference between the number of observation equations and the number of unknowns.

The theoretical precision of the estimated parameters and coordinates can be obtained from the covariance matrix:

$$\mathbf{C}_{\mathbf{xx}} = \sigma_0^2\mathbf{N}^{-1} \tag{11}$$

An alternative method to set up the reference system is the use of a set of well-distributed and sufficiently precise ground CP's (GCP's—see Smith and Atkinson 2001). In such case, GCP coordinates may be used in Eq. (3) instead of the feature coordinates on the master. This option is expected to work well if all GCP's have a uniform precision, much better than the one needed for the co-registration process. A second option is to build up a system including only a set of Eqs. (4) for the slave-to-slave correspondences (in this case, all images would be treated as slaves), complemented by some *pseudo-observation* equations where the GCP coordinates are fixed to the known values. This approach also enables to deal with GCP's having different precisions.

3 Case Study

3.1 Study Area and Data

The global registration approach described in Sect. 2 was used for the automatic co-registration of a Landsat time series collected over Valtellina (Rhaetian Alps in northern Italy), where a flood and the consequent landslides occurred in 1987.

The emergency in Valtellina started in the second half of July 1987 and lasted until the beginning of September 1987. Heavy rainfall and the melting of glaciers resulted in the flooding of several rivers and streams which caused heavy damages to villages and towns. For more details, the reader is referred to Alexander (1988).

On Tuesday, 28th July 1987, a large landslide occurred in Val Pola (Province of Sondrio, northern Italy). More than 40×10^6 m^3 of rock debris came off from the flank of Monte Zandila moving downslope, reaching the bottom of the valley and partially the opposite side, see Figs. 1 and 2. Some villages were destroyed resulting in heavy casualties (Azzoni et al. 1992; Crosta et al. 2004). The bulk of rock and debris created an artificial dam by halting the flow of the River Adda, as well as interrupting the main road to the upper Valtellina valley. Large efforts and costs were needed for recovering the transportation network and for draining the lake, whose volume reached a maximum of about 6×10^6 m^3 of water.

A satellite time series made up of thirteen medium-resolution Landsat-4 and Landsat-5 Thematic Mapper (TM) images (path 197, row 28), collected from 1984 to 2011, was selected for this study. The list of the adopted images is reported in Table 1. The data were made available by the Earth Resources Observation and Science Center (EROS) data center of the US Geological Survey (USGS) and included scenes imaged in different seasons and years, with different cloud fraction, snow cover, and more generally with significant land-cover changes. Moreover, the images taken in 1987 had a high cloud cover, making their processing quite complex.

Fig. 1 A picture of the Val Pola landslide and the lake formed by the dam of debris. *Source* www.gearthhacks.com

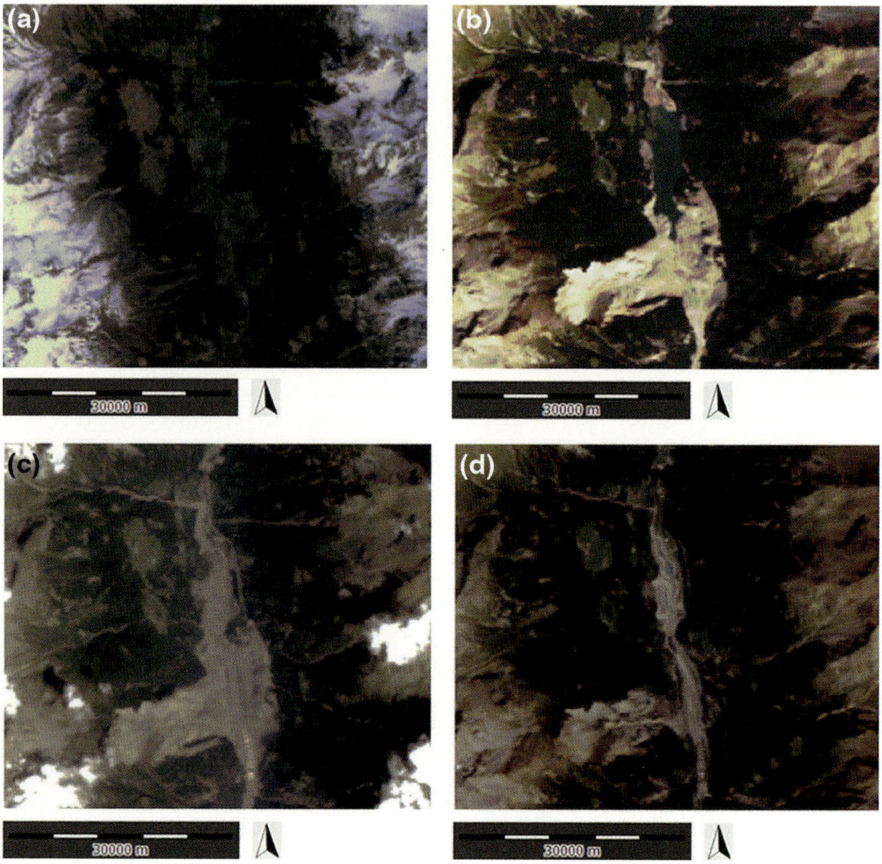

Fig. 2 Satellite images of the study area collected: **a** before the landslide (29th April 1987); **b** after the landslide (20th September 1987); **c** after the drainage of the artificial lake (3rd August 1990); **d** 14 years after the landslide (21st August 2011)

3.2 Data Processing

All the Landsat/TM full frames were processed at their original resolution, without any geometric down-sampling or geographic tiling. Spectral band TM3 (0.63–0.69 μm) was used for the FBM. After image matching, a *connection graph* was created (Fig. 3) for highlighting all the links between the images composing the time series. It is worth to mention that in this case study the connection graph clustered the data into two main blocks: Image #1 had several connections with images #6–13 because they were all collected in summer (from July to September), while images #2–5, collected in winter (from January to April), were grouped in a different cluster. In any case, all the clusters were connected together, and thus, a global adjustment of the whole time series was possible.

Table 1 Landsat/TM data set used in the study

ID	Image source	Acquisition date	Satellite/ sensor	Ground resolution (m)
1	LT51930281984207AAA04	25th July 1984	Landsat-5/TM	30
2(*)	LT51930281987007XXX01	7th January 1987	Landsat-5/TM	30
3(*)	LT51930281987023XXX01	23rd January 1987	Landsat-5/TM	30
4(*)	LT51930281987039XXX01	8th February 1987	Landsat-5/TM	30
5(*)	LT51930281987119XXX02	29th April 1987	Landsat-5/TM	30
6(*)	LT51930281987183XXX02	2nd July 1987	Landsat-5/TM	30
7(*)	LT51930281987215XXX02	3rd August 1987	Landsat-5/TM	30
8(*)	LT51930281987263XXX02	20th September 1987	Landsat-5/TM	30
9	LT41930281988226XXX05	13th August 1988	Landsat-4/TM	30
10	LT41930281990215XXX03	3rd August 1990	Landsat-4/TM	30
11	LT51930282003259MTI01	16th September 2003	Landsat-5/TM	30
12	LT51930282007238MOR00	26th August 2007	Landsat-5/TM	30
13	LT51930282011233MOR00	21st August 2011	Landsat-5/TM	30

(*) Images collected in the year of Val Pola landslide

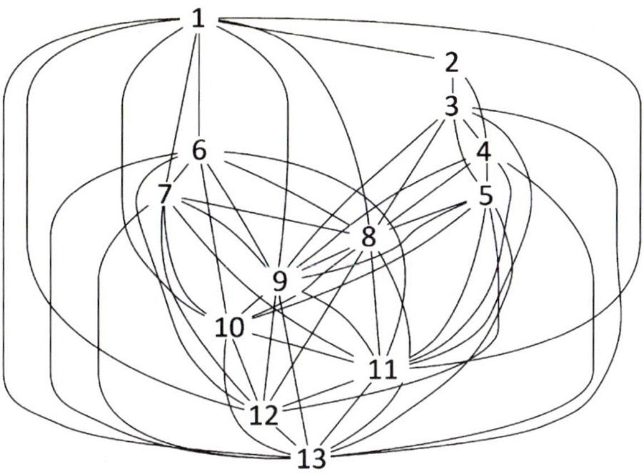

Fig. 3 Connection graph of the satellite time series showing how data are related and clustered

It is important to notice that different images may have different number of features with different spatial distribution, as shown in Fig. 4. For instance, the images acquired in summer and not during the emergency had many features, whereas those acquired during the year of the flood were characterized by fewer features because of an overall higher cloud cover. From this point of view, the method is also able to highlight the presence of clouds or other anomalies in the data set.

As already mentioned, all the images were included in the adjustment because the connection graph suggested that this operation was feasible using a system of equations combining both categories of Eqs. (3) and (4). In addition, the algorithm automatically discarded the image pairs with fewer corresponding features: A minimum

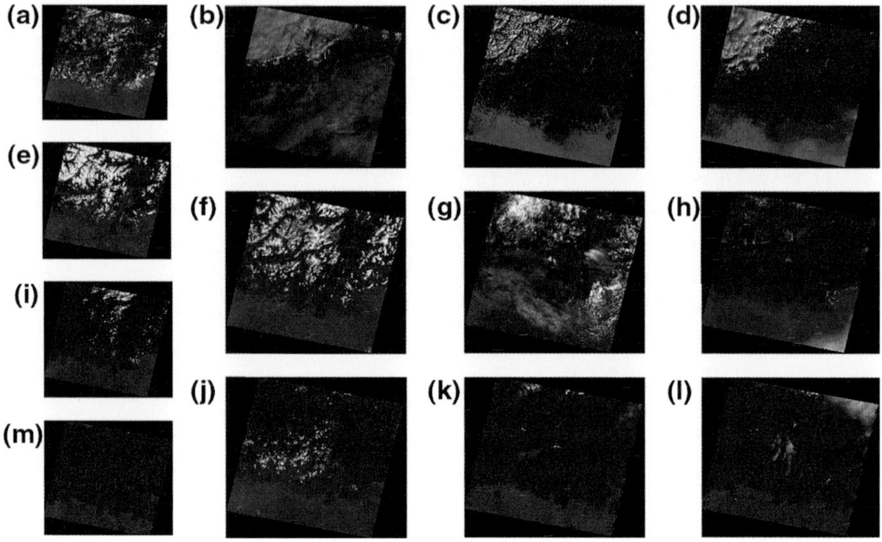

Fig. 4 Matched features overlaid to the Landsat/TM images. It is clear that images with a high snow and/or cloud cover have an uneven feature pattern. The data set includes the following images: **a** 25th July 1984, **b** 7th January 1987, **c** 23rd January 1987, **d** 8th February 1987, **e** 29th April 1987, **f** 2nd July 1987, **g** 3rd August 1987, **h** 20th September 1987, **i** 13th August 1988, **j** 3rd August 1990, **k** 16th September 2003, **l** 26th August 2007, **m** 21st August 2011

threshold of 40 features per image was selected for the experiments. This additional condition was included to avoid the use of images whose co-registration could not be accomplished in a reliable way, i.e., based on a too small number of corresponding observations.

Finally, results were compared with a manual co-registration of the whole time series done by an expert operator.

4 Results and Discussion

Subpixel precision was achieved with both manual registration and automatic data processing (see Table 2), notwithstanding the manual stacking yielded slightly better results as witnessed by the smaller value of the estimated sigma naught (0.57 pixels for the manual and 0.73 pixels for the automatic processing). Indeed, the sigma naught of Least-Squares gives a picture of the average estimated precision of the adjusted observations.

The computed parameters show that Landsat/TM images were essentially affected by different translations, while the scale factors s_j were almost unary and rotations α_j were null. Results were very similar for both manual and automatic measurements. A synthesis is given by means and standard deviations of the scale factors and rotation angles shown in Table 3.

Table 2 Summary of the Least-squares adjustment

	Automatic measurement	Manual measurement
Equations (No.)	59,720	922
Unknowns (No.)	25,968	78
Sigma naught (pixel)	0.73	0.57

Table 3 Mean and standard deviation for scale factors and rotation angles

	Automatic measurement	Manual measurement
$\overline{s}$	1.000225	1.000193
σ_s	0.000069	0.000027
$\overline{\alpha}$ (°)	0.00192	0.00089
σ_α (°)	0.00147	0.00062

The translation vectors were instead significantly different for every image, confirming the assumption illustrated in Sect. 2.1. A comparison between manual and automatic measurements $\left(\Delta t_x = t_{x\_man} - t_{x\_auto} \text{ and } \Delta t_y = t_{y\_man} - t_{y\_auto}\right)$ was carried out to define the accuracy of data registration. The means of these differences, for x and y components, were −0.05 and −0.10 pixels, whereas the standard deviations were 0.36 and 0.35 pixels, respectively. This confirmed that both the manual and automatic results were consistent and that subpixel accuracy was reached during the registration phase.

Finally, a consideration about the variance–covariance matrix $\mathbf{C_{xx}}$ of the estimated parameters deserves to be mentioned. Given a generic image j, the variances (diagonal elements of $\mathbf{C_{xx}}$) of the translation parameters in both directions x and y are the same $\left(\sigma^2_{tx,i} = \sigma^2_{ty,i}\right)$. Obviously, different images have different variances, depending on the number of features, their spatial pattern, and the connections to other images. The average theoretical precision of the translation parameters (extracted from the diagonal elements of $\mathbf{C_{xx}}$) is given by the root mean squares (RMS):

$$\overline{RMS_{tx}} = \sqrt{\sum_{i=2}^{13} \frac{\sigma^2_{tx,i}}{N-1}} \tag{12}$$

In this case, $\overline{RMS_{tx}}$ resulted in 0.17 pixels for the automatic measurements and 0.19 pixels for the manual registration. These values are comparable with the ones previously obtained concerning the accuracy evaluation of the translation parameters. When comparing these results with traditional state-of-the-art techniques, the advantage of the multi-image registration described in this chapter concerns the simultaneous estimation of all the transformation parameters with increased reliability. The use of a FBM technique based on SURF operator makes this method fully automatic and very robust against outliers, as gross and small error detection are included after the matching of every image pair combination and, eventually, during the final Least-squares adjustment. Consequently, the accuracy of the full automatic data processing (without human interaction) is similar to that of the manual co-registration.

The proposed method also avoids standard image matching between several slave images and only one master. Thus, also images of the time series that do not share any corresponding features with the master, but do have some correspondences with one or more other slaves, can be registered as well. The case study with medium-resolution satellite data proved that manual measurements are not needed, notwithstanding the complexity of the land cover and its changes over time, and different snow and cloud cover and modifications of the landscape due to the flood and landslide. However, two main computational issues are still open: (i) the matching time that depends on N^2 and (ii) the size of the normal matrix (6) in the Least-squares problem. The first matter could be partially overcome with speeded-up matching strategies for the comparison of descriptors, whereas after image matching the features may be reduced by considering their spatial pattern in the images and their multiplicity. Moreover, the computational power of CPU's is continuously increasing and code parallelization or implementations that exploit the power of modern GPU's can significantly reduce the computational time. Regarding the second matter, the size of the normal matrix mainly depends on the number of extracted features, whereas the number of images of the data set has only a limited influence. For this reason, a strategy able to decimate spatial features before running the adjustment algorithm could be a valid solution to the second problem.

5 Final Considerations

This chapter describes a new approach to full automatic image registration of satellite time series by means of a simultaneous adjustment. The method allows one to overcome some intrinsic limits of standard image-to-image registration techniques, mainly the low reliability and the weakness against potential blunders.

The developed methodology first extracts corresponding features in a pairwise manner by considering any image pair combinations. A feature-based matching (FBM) technique based on SURF operator is applied to this purpose. After recognition and reordering of multiple features (that can be tracked over more than two images), all resulting observations (i.e., feature point coordinates) are included into a redundant system of observations, which can be solved via Least-squares adjustment. The output is given by the parameters of the transformations to map all the images with respect to a reference one ('master'). Thanks to the extraction of corresponding features between any image pairs, the images that do not directly share corresponding features with the master can be also included into the global adjustment.

Here, the application of this methodology was focused on medium-resolution satellite data. Results in terms of average registration precision were found at subpixel level; thus, a 2D similarity transformation seems enough for most of the applications involving the use of medium-resolution satellite imagery. The step after image co-registration is the extraction of information useful for specific landslide investigations. Several authors proposed the use of medium-resolution satellite images, such as Landsat, for monitoring the dynamics of landslides over time (Metternich et al. 2005; Guzzetti et al. 2012). For example, the visible and near-infrared spectral bands, typically recorded by medium-resolution multispectral sensors, can be useful for the calculation of vegetation indexes (Yang et al. 2012; Qiao et al. 2013). These parameters are useful to recognize the areas where vegetation has decreased or disappeared, which may be likely related to landslides. With respect to the measurement of surface displacements, optical-flow tracking (Baker et al. 2011) or other image matching techniques (Scaioni et al. 2014b) can be used only if the displacements being measured between different epochs are comparable to the pixel size of the images used, thus few dozens of meters in the case of medium-resolution satellite data. In addition, the same co-registration techniques could be used in related research fields, like investigation on the ice-flow velocity of glaciers, where the displacements may be quite large for being well recognized with medium-resolution images, see Heid and Kääb (2012).

It is worth to highlight an important property of the multi-image registration method addressed in this chapter, which is particularly relevant for landslide monitoring applications: The robustness against blunders. Indeed, when the image content changes due to the occurrence of a landslide, the FBM process is not able to output any features in the area of such disruptive event. On the other hand, if a landslide body is steadily moving, but it is still far from the final run-out, the effect of FBM may be different. If a portion of the surface rigidly shifts between the acquisitions of two images, those corresponding features located on the landslide will not follow the global mapping model that works out well for the stable areas. Under a statistical point of view, these features are considered as outliers and have to be removed. The use of a long series of images along with a multi-image registration method has more chance to identify the features on stable areas. By looking at the connection graph (see the example in Fig. 3), features which have been identified only for limited laps are supposed to be in unstable areas and discarded. On the other hand, features that are connected to most images of the series might probably be in a moving region.

Acknowledgments This work has been supported by the Italian Ministry of Education, University and Research (MIUR) within the grant FIRB 2010 entitled: 'Subpixel techniques for matching, image registration and change detection with applications to civil and environmental engineering' (No. RBFR10NM3Z). Acknowledgements also go to the 973 National Basic Research Program of China 973 (No. 2013CB733204).

References

Alexander, D. (1988). Valtellina landslide and flood emergency, Northern Italy, 1987. *Disasters, 12*(3), 212–222.

Apollonio, F. I., Ballabeni, A., Gaiani, M., & Remondino, F. (2014). Evaluation of feature-based methods for automated network orientation. *International Archives of the Photogrammetry Remote Sensing and Spatial Information Sciences, 40*(5), 47–54.

Azzoni, A., Chiesa, S., Frassoni, A., & Govi, M. (1992). The Valpola landslide. *Engineering Geology, 33*(1), 59–70.

Bay, H., Ess, A., Tuytelaars, T., & Van Gool, L. (2008). SURF: Speeded up robust features. *Computer Vision and Image Understanding, 110*(3), 346–359.

Baker, S., Scharstein, D., Lewis, J. P., Roth, S., Black, M. J., & Szeliski, R. (2011). A database and evaluation methodology for optical flow. *International Journal of Computer Vision, 92*(1), 1–31.

Barazzetti, L., Remondino, F., & Scaioni, M. (2010). Orientation and 3D modelling from markerless terrestrial images: Combining accuracy with automation. *The Photogrammetric Record, 25*(132), 356–381.

Barazzetti, L., Gianinetto, M., Scaioni, M. (2014a). Simultaneous least-squares registration of satellite time series. In: *Proceedings of IEEE International Conference WHISPERS 2014* (p. 4). June 24–27 2014, Lausanne, Switzerland.

Barazzetti, L., Gianinetto, M., & Scaioni, M. (2014b). Automatic co-registration of satellite time series via least squares adjustment. *European Journal of Remote Sensing, 47*, 55–74.

Bouchiha, R., & Besbes, K. (2013). Automatic remote-sensing image registration using SURF. *International Journal of Computer Theory and Engineering, 5*(1), 88–92.

Brown Gottesfeld, L. (1992). A survey of image registration techniques. *ACM Computing Surveys, 24*, 325–376.

Castro, E. D., & Morandi, C. (1987). Registration of translated and rotated images using finite Fourier transform. *IEEE Transactions on Pattern Analysis and Machine Intelligence, 9*, 700–703.

Crosta, G. B., Chen, H., & Lee, C. F. (2004). Replay of the 1987 Val Pola Landslide, Italian Alps. *Geomorphology, 69*, 127–146.

Dowman, I., & Tao, V. (2002). An update on the use of rational functions for photogrammetric restitution. *ISPRS Journal of Photogrammetry and Remote Sensing, 7*(3), 26–29.

Fiorucci, F., Cardinali, M., Carlá, R., Rossi, M., Mondini, A. C., Santurri, L., Ardizzone, F., & Guzzetti, F. (2011). Seasonal landslide mapping and estimation of landslide mobilization rates using aerial and satellite images. *Geomorphology, 129*(1–2), 50–70.

Fischler, M. A., & Bolles, R. C. (1981). Random sample consensus: A paradigm for model fitting with applications to image analysis and automated cartography. *Communications of the ACM, 24*(6), 381–395.

Gianinetto, M., & Scaioni, M. (2008). Automated geometric correction of high-resolution pushbroom satellite data. *Photogrammetric Engineering and Remote Sensing, 74*(1), 107–116.

Gianinetto, M. (2012). Automatic co-registration of satellite time series. *The Photogrammetric Record, 27*(140), 462–470.

Goshtasby, A., Stockman, G. C., & Page, C. V. (1986). A region-based approach to digital image registration with subpixel accuracy. *IEEE Transactions on Geoscience and Remote Sensing, 24*, 390–399.

Gosthtasby, A. (2005). *2-D and 3-D image registration* (p. 258). Hoboken, NJ, USA: Wiley.

Grün, A. (2012). Development and status of image matching in photogrammetry. *The Photogrammetric Record, 27*, 36–57.

Guzzetti, F., Mondini, A. C., Cardinali, M., Fiorucci, F., Santangelo, M., & Chang, K. T. (2012). Landslide inventory maps: New tools for an old problem. *Earth-Science Reviews, 112*, 42–66.

Heid, T., & Kääb, A. (2012). Evaluation of existing image matching methods for deriving glacier surface displacements globally from optical satellite imagery. *Remote Sensing of Environment, 118*, 339–355.

Khairunizza-Bejo, S., & Petrou, M. (2010). Elastic image registration for landslide monitoring. *International Journal of Signal Processing, 3*(3), 71–86.

Kraus, K. (1997). *Photogrammetry. Volume 2. Advanced methods and applications* (p. 466). Bonn: Dümmler Verlag.

Kraus, K. (2008). *Photogrammetry: Geometry from images and laser scans* (p. 459). Germany: Walter de Gruyter.

Le Moigne, J., Netanyahu, N. S., & Eastman, R. D. (2011). *Image registration for remote sensing* (p. 484). UK: Cambridge University Press.

Maianti, P., Rusmini, M., Tortini, R., Dalla Via, G., Frassy, F., Marchesi, A., Rota Nodari, F., & Gianinetto, M. (2014). Monitoring large oil slick dynamics with moderate resolution multispectral satellite data. *Natural Hazards, 73*, 473–492.

Mantovani, F., Soeters, R., & Van Westen, C. J. (1996). Remote sensing techniques for landslide studies and hazard zonation in Europe. *Geomorphology, 15*(2), 213–225.

Metternicht, G., Hurni, L., & Gogu, R. (2005). Remote sensing of landslides: An analysis of the potential contribution to geo-spatial systems for hazard assessment in mountainous environments. *Remote Sensing of Environment, 98*, 284–303.

Mikhail, E. M., Bethel, J. S., & McGlone, J. C. (2001). *Introduction to modern photogrammetry* (p. 479). New York: Wiley.

Nayak, S., & Zlatanova, S. (2008). *Remote sensing and GIS technologies for monitoring and prediction of disasters* (p. 271). Berlin, Heidelberg: Springer.

Poli, D., & Toutin, T. (2012). Review of developments in geometric modelling for high resolution satellite pushbroom sensors. *The Photogrammetric Record, 27*(137), 58–73.

Pluim, J. P. W., Maintz, J. B. A., & Viergever, M. A. (2001). Mutual information matching in multiresolution contexts. *Image and Vision Computing, 19*, 45–52.

Pratt, W. K. (1991). *Digital image processing* (2nd ed.). New York: Wiley.

Price, K. E. (1985). Relaxation matching techniques—a comparison. *IEEE Transactions on Pattern Analysis and Machine Intelligence, 7*, 617–623.

Qiao, G., Lu, P., Scaioni, M., Xu, S., Tong, X., Feng, T., et al. (2013). Landslide investigation with remote sensing and sensor network: From susceptibility mapping and scaled-down simulation towards in situ sensor network design. *Remote Sensing, 5*(9), 4319–4346.

Scaioni, M., Longoni, L., Melillo, V., & Papini, M. (2014a). Remote sensing for landslide investigations: An overview on recent achievements and perspectives. *Remote Sensing, 6*(10), 9600–9652. doi:10.3390/rs6109600.

Scaioni, M., Feng, T., Lu, P., Qiao, G., Tong, X., Li, R., et al. (2014b). Close-range photogrammetric techniques for deformation measurement: Applications to landslides. In M. Scaioni (Ed.), *Modern technologies for landslide investigation and prediction* (pp. 13–41). Berlin, Heidelberg: Springer.

Smith, D. P., & Atkinson, S. F. (2001). Accuracy of rectification using topographic map versus ground control points. *Photogrammetric Engineering and Remote Sensing, 67*(5), 565–570.

Teke, M., & Temizel, A. (2010). Multi-spectral Satellite image registration using scale-restricted SURF. In: *Proceedings of 20th International Conference on Pattern Recognition* (pp. 2310–2313). August 23–26 2010, Istanbul, Turkey.

Wahed, M., El-tawel, G. S., & Gad El-karim, A. (2013). Automatic image registration technique of remote sensing images. *International Journal of Advanced Computer Science and Applications, 4*(2), 177–187.

Yang, C. J. L., Ren, X. L., & Huang, H. (2012). The vegetation damage assessment of the Wenchuan earthquake of May 2008 using remote sensing and GIS. *Natural Hazards, 62*, 45–55.

Zlatanova, S., & Li, J. (2008). *Geospatial information technology for emergency response* (p. 381). London: Taylor & Francis Group.

Printed by Printforce, the Netherlands